Endorsements

Andy Fletcher is a gifted presenter who I would welcome into the Seoul Foreign School TOK class without hesitation. Through his presentations he challenges IB students to contemplate the nature of knowledge and of our world, focusing on the very core of the focus of a theory of knowledge course. As a (former) teacher at international schools he fully understands both international students and the value of the Theory of Knowledge course. He has presented at international school conferences and been invited into numerous TOK classrooms in the US and throughout the world. I believe that he would add meaningfully to the TOK course taught at any international school and recommend him unreservedly. He will stimulate and challenge your students' perceptions and beliefs in ways that enhance the effectiveness of what I believe to be one of the most important components of the IB program, the Theory of Knowledge.

- Harlan E. Lyso, Head of School, Seoul Foreign School

I have just come from one of Andy's talks on the links between science and religion. It was entertaining, engaging and very funny - but better than this it was inspirational. Whether or not you believe that science can tell us anything about religion - (I should be frank here and say that I do not) this is the perfect way to bring students into the problem and make it immediate and real for them. All the students were fascinated by his mesmerising presentation and left wanting to know more. I heard them talking about it at lunchtime; they were enthused, intrigued and in some cases enraged - but they all wanted to talk about it, to explore the ideas and to see how their beliefs stood up to the themes he presented. This was a terrific experience for the kids - even though I violently disagree with some of Andy's premises, bringing up the ideas in the way he does is, I believe, what education should be all about. Andy is a wonderful guest speaker; I am moving schools this year and I will do my best to bring him to my new school - and I would recommend him to anyone wishing to explore the links between science and religion.

- Nick Alchin, author *Theory of Knowledge* (Pupils' book and Teachers' Guide), National Technological University (Singapore), California State University (Singapore campus), United World College of SE Asia (Singapore), International School of Geneva (now at Sevenoaks School, Kent, UK); March 2003

Having just spent a week with (Andy Fletcher), I feel compelled to write and express my utter delight and gratification for his efforts in my high school classroom. (He) guest-lectured in my classroom for four days last week. It was a daunting task. His topics were culled from post-modern science, including Einsteinian relativity, quantum mechanics,

1

chaos theory, and complexity theory. Andy's knowledge base for these complex and ill-understood science topics is incredible. Not only has he command of the subject matter, but he has the ability, as the gifted teacher he is, to relate the information in a fun, informative, and, most importantly, relational manner. While he is judicious, as he should be, about overtly referencing issues of faith, he weaves a compelling story about how the topics of 20th century science can lead to rigorous self-questioning about creation, religion, and one's place in the universe. I have taught for twenty-two years and received several teaching awards myself, so I recognize quality teaching when I see it. Andy was absolutely spell-binding. I found myself in awe of this master teacher taking my students on a most exciting and worthwhile journey -- Andy created some magic moments in my classroom.

- Dr. John Kistler, Palmer High School, Colorado; January 2002

This is a very brief note to say very hearty and genuine thanks to you for your visit to this strange part of the world. From our point of view it was a real highlight: educational, enjoyable, stimulating and provocative and will, I know, have made a lasting mark on our lads and lasses. Neil and James have already been using the DVDs in TOK and Mark has asked to use them for his biology: naturally I will be drawing upon material in them for my physics and chem. too. I particularly want to use the chaos one with my year 10 class to stretch their minds a little beyond the mundanities of IGCSE Coordinated Science

- Glynn Morgan, Academia Británica Cuscatleca, Santa Tecla, El Salvador, November 2003

I have received parent feedback re your presentation here at Ridgeview in February. The consensus was that they've never seen their kids so animated about what went on in school. The feedback from both juniors and seniors was extremely favorable, except for the all-day session and lack of AC in the library. One young lady's comment summed it up for all: "Biology is my science area. For me to listen to 6 hours of physics and not be bored -- it had to be awesome!" They have asked to have you back again next year. I am writing your presentation into my budget for next year.

- Vickie Bandy, Ridgeview High School, Orange Park Florida USA, February 2005

Many thanks for the stimulating seminars. I am sure the students who come to school today were not expecting to be so mentally challenged and fascinated on their last day of school for the term.

- Mike Rice, United World College of South East Asia, Singapore, December 2004

He made listeners stop and think. Whether or not they believe that religion does have a place in science doesn't matter. He made the kids think and question themselves about their own beliefs. My students did not know what to think about his presentations. They thought that they would be dull, boring and uninteresting. Instead, they were amazed that they were enthused and intrigued and in some cases, enraged about the ideas and concepts in his presentations. His ability to bring the complex theories down to our level was amazing. I can see that he was a gifted teacher because he has the ability to relay the information in a fun, informative and relational manner while also being entertaining.

- Bettye Francis, IB Program/ADP Coordinator, Mt. St. Mary Academy, Little Rock AR USA, 2004

The presentation was bound to be controversial given the current state of affairs and the subject: science and religion. (He) took us from the assumptions of Newton's theory to Einstein, quantum physics, chaos theory, the butterfly effect and complexity in an easy-to-understand fashion...Students talked about the presentation for some time. On the day, we had already extended the session by another forty minutes but some students stayed for a further forty! The seminars made our students think and reminded them that learning and thinking are both interesting and provocative...It was a great pleasure to note that a number of parents also attended and that AP and other students for whom the seminars were not mandatory were there for the full sessions. All in all, this was a most successful venture for ACS Hillingdon.

- Bernadette Clare, ACS Hillingdon International School, UK, November 2005

STUDENTS

Thank you for your time and for your ideas. The presentation was the best thing I've been taught since I started school. - Brett, September 2002

Your presentation has changed the way I see the world. Thank you. - Chazalyn, September 2003

The last five days were without a doubt one of the best presentations ever in my life! - Sean, September 2002

Thank you. Your presentation was truly life changing. - Paloma, September 2003

"Awesome in the true sense of the word"; "Accessible, funny and interesting"; "He made us think and learn in an exciting way." -- ACS Hillingdon International School (UK) students, November 2005

Andy Fletcher's presentation was a lot of information to take in on one day. Nevertheless, it proved to be a presentation about fascinating facts we don't hear every day. It grabbed my attention from the beginning and held it till the very end. It provided evidence and possible answers to questions that a lot of people are searching for in this *day and age. For instance, a lot of people wonder when and how the universe started, and "why are we here?" Some of the facts were almost impossible to believe, while other facts were just too fascinating for words. The topic that interested me most was the different theories about the universe (from) Newton , Einstein etc. It has inspired me and will be a great help when writing my final draft for my TOK essay. It will help me answer questions people ask me about religion. I would love to read Fletcher's book, because I would love to know more about "Life, the universe and everything". I also believe it's a good thing to "listen" to other's ideas with an open mind. Even if we don't agree with them, we might just learn something from what they say, and this will help us and understand or criticise our own views with more insight. It could also teach us tolerance towards different people's opinions.* -- Berlene, Westwood IS Botswana, June 2006*

Going into the hall on that Friday morning, I was not too enthusiastic about the hours to come. I was defensive, as I always am about this sort of thing where everything is questioned. But as soon as Andy began to talk, I mentally relaxed as he threw humour into the mix and began to baffle our minds with all sorts of ideas and facts that we had never heard of before. What truly got me more willing to listen though was the way that he began. The idea(s) all completely threw my perception of the world out the window. There is so much out there that we still do not understand and will never fully comprehend. This led to the theory of the Big Bang which I have been against my whole life. Overall, I was amazed at the way that these new ideas had broadened my perceptions to such a large extent. My faith was not attacked and I came out questioning and wanting to know more about the unknown world around me. Time, fractals, butterfly effects and mere existence itself have all got a lot to offer for the scientists who have only seen the tip of the iceberg. I want to know more. -- Kira, Westwood IS Botswana, June 2006

Life, the Universe, and Everything

Investigating God and the New Physics

Life, the Universe, and Everything

Investigating God and the New Physics

Andy Fletcher

Published by **fletchpub** at Lulu
Additional copies available at
www.lulu.com/content/146094

Front Cover Design by Lulu
Back Cover Photo from NASA
ISBN: 1-4116-7369-7
Set in 12 point Garamond type

The subject matter of this book is taken from a series of
seminars developed and given worldwide by Andy Fletcher.
For information on hosting the seminars, to see the
bibliography from which the seminars were developed, to see
a list of places where the seminars have been given, to buy
the seminars on DVD or VHS, or for a history of the
seminars and the organization, please visit
www.tokseminars.org.

Life, the Universe, and Everything, Inc.
Also known as TOK Seminars
c/o TOK Seminars, PO Box 104,
Monument CO 80132 USA
Andy Fletcher, President and Executive Director
001.719.282.8783

Table of Contents

Preface

Let's just say that first of all, this is not a book for scientists.

(Scientist)

(Non-scientist)

There are a lot of books already that scientists wrote for the common reader, and although most of them are pretty good, they really are written for dumb scientists rather than for smart normal people.

I know, because I've read them all. Mostly. The only way I survived was to learn how to skip all the parts that weren't written for me but for all the scientist's friends. That meant that for some books and writers (and you know who you are), I pretty much just read the first paragraph or two and the last page of each chapter. I would ask for a refund for all the unused pages, but I bought them all in used-book stores anyway. Besides, now I've got a very intimidating library of books that look well-used (because somebody else well-used them before I bought them) that I can impress all my friends with. That's key to creating the impression of great intelligence—have a lot of weighty-looking books that you know enough about to talk to, say, Oprah, but not to, say, Stephen Hawking.

So here's this book about physics and God and so on, taken from a series of seminars by the same title that I do in public high schools and international schools. I have to confess that I stole the title completely.

Life, the Universe, and Everything comes from Douglas Adams' *Hitchhiker's Guide to the Galaxy* (the best five-book trilogy ever written), and I use it because I love the *Guide,* even though hardly anybody reads it anymore. Plus there's the part about the old guy who used Chaos Theory to make the coastlines of Norway, which I used to include in my seminars, but don't anymore.

I stole the last part of the title from Paul Davies, a quantum gravity specialist and "Professor of Natural Philosophy in the Australian Centre for Astrobiology at Macquarie University," as it says on his website, among a lot of other things. One of his many great books (I read them all, which is to say, I read some of each them) is called *God and the New Physics.*

I stole the only apparently original word, *Investigating,* from my friend Dr. Paul Burkimsher at CERN in Geneva, who suggested it.

So now it's time for the real confessional (see how the God parts keep sneaking in?). I'm not a scientist. I could have said, like Dr. Science, that I'm not a real scientist, but I'm not even a pretend scientist. I squeaked through Mr. Stillman's biology class by the (to use a biological term) skin of my teeth. I did great in Mr. Deleay's chemistry class, but Mr. Kinniburgh couldn't tempt me into physics. I took some of a physics class in college, mainly because of this really cute girl named Vickie who was in the class, but then my dad showed up one day (he lived a thousand miles away), having heard somehow (this was before email, but after the telephone) that I was spending more time on flirting than on physics, so I dropped the course. This is true.

Yep. Dropped the only physics course I ever had. OK, truth in advertising: I took almost everything you need for a math degree, but got sidetracked somehow into history, but then I taught high school and junior college math for a long time, so I'm not a total loser, except insofar as math geeks are inherently losers, but that's beside the point.

Thing is, basic physics is (how to put this nicely) a little dull, sorta pedestrian, kinda (to foreshadow this book) mechanical and reductionistic. Uninteresting. Not sexy the way high energy particle physics and cosmology are sexy. I mean, I guess we have to have the dull stuff so that clocks and cars will work, but this other stuff is way cool.

So this is a book about physics. It's not a book about God, except when God shows up in a physics sort of way. In fact, it's a book about the really disconcerting way that God keeps showing up in a physics sort of way. All the physicists are talking about it.

That's why we have articles every now and then in magazines and newspapers that talk about it, like *Newsweek*'s July 1998 cover, "Science Finds God," and the *Saturday Evening Post* with the same headline in January 1999. *Scientific American*'s website had an article called "Beyond Physics: Renowned scientists contemplate the evidence for God." *Wired* magazine had a cover entitled "Science Gets Religion" in December 2002. The *New York Times* ran a story several years ago by real scientist James Gleick, who quoted this from John Updike's novel *Roger's Version:* "The most miraculous thing is happening. The physicists are getting down to the nitty-gritty; they've really just about pared things down to the ultimate details, and the last thing that anybody ever expected to happen is happening. God is showing through."[1]

The physicists can't quite make up their minds about God and physics, so one day they lean one way, the next day they lean another. Truth is, they are all sort of hoping that they'll make some discoveries that will take God out of physics and put him back into church, where they think he belongs. They keep trying, which is what physicists do—it's their job. They've got the multi-verse theory, infinite universes theory, string theory, M-theory (ask them what the M stands for; they don't know and it drives them crazy), and those theories are all kinda cool, but none of them are science yet – there's no evidence for them to be found anywhere yet, and there may never be. This is a book about mainstream, everybody-believes-it, this-is-what-they-teach-in-school science. The science in this book is weird, but that's not my fault. But it's not *National Enquirer,* Elvis/Bigfoot/Area 51/Space Alien science, even though it might sound like it every now and then.

So, there you go. Read the book. Buy it first. I need the money. If you're buying it in a used-book store, send me a dollar, just so that I know someone's reading it. Plus, I need the money.

Questions

1. Why should you buy this book?

      a) I need the money

      b) All of the above

1

Dated Ideas: How We Thought the Universe Works

The universe is a lot more interesting than you think it is. That's true even if you think it's pretty interesting, because when I say "interesting," I don't mean beautiful, awesome, unbelievably huge, spectacular, or amazing, though it is all of those things, too. When I say "interesting," I mean that learning more about the universe is going to rock your socks, blow your doors off, and change the way you think about everything. There are only a couple of ways that won't happen: if (1) you stop reading right now; or (2) you already know all this stuff. Which most people don't. And when I say "most," I mean "all."

You'll have to wait a bit, though.

Scientists Discover the Universe

Once upon a time there was a man named Isaac Newton. He was an extraordinarily smart guy, smart enough that he looked around at nature and the universe and decided that something was missing, so he invented calculus to describe it.

Isaac Newton as an '80s rock musician named Peter Frampton

I took calculus. I was really, really good at calculus (it was pretty much the last thing in math I was really good at). As a paid math tutor I even taught people who were ahead of me in the four-course calculus series.

Then I took the theory of calculus. The difference in *being good* at doing calculus and *understanding* how calculus works is the same difference between knowing how to *drive* a car, and knowing how to *build* a car starting with Legos, raw coal, and rubber trees. Newton was a bright guy.

He and his contemporaries (this was back in the 1600s) generally saw the universe as what can be described in four simple terms: it was infinite in space and time, deterministic, reductionistic, and mechanistic (IDRM, for short). We'll define those as we go along.

What they really meant was that the universe was like a big clock, except that since it was infinitely old, nobody needed to wind it up and give the pendulum a push in the first place. It was perpetually wound, a clock that never stopped ticking, and more importantly, never started ticking. It has always been ticking.

Infinite and Mechanistic

It worked like a machine, and since machines were pretty much brand new in those days, that is, the first machines were being invented at the time, and everyone was so impressed with machines (they just thought machines were wonderful, which they are), it seemed logical that the universe would be like a machine. Things like steam turbines, pendulum clocks, and the candy cane (which is not a machine, but is nevertheless an impressive invention) were being trotted out by the dozens and impressing everyone in sight, so let's trot out the universe and show that it's a machine, too.

That's what an infinite, mechanistic universe is—a clock that never stops ticking, and never started, either, since it's been going forever. Two things to remember about this idea of an infinite, mechanistic

18

universe: The universe is infinite (duh) and goes out forever in space, back forever in time. And it's predictable, like machines are predictable. So you can always tell what's going to happen in nature by coming up with scientific descriptions of what's happened in the past and predicting on that basis what will happen in the future.

And like a machine, they assumed, mostly accurately but not completely, that you could take the universe apart to figure out what made it work. The smaller the parts, the better you understand how it works.

Reductionistic

That's what a reductionistic universe is. You can "reduce" it to its smallest parts to find out how it works, and the smaller the parts, the better you know. That also means that nothing in nature is anything more than just the sum of its parts. A car is nothing more than car parts. A plane is nothing more than plane parts. You are nothing more than you parts. The universe is nothing more than the parts of the universe, all stuck together to look just like a universe, just like your you parts are all stuck together to look just like you.

Back in my wild and misspent youth, just after Newton died, Mr. Stillman, my biology teacher, used to reduce restaurant frogs for us. (I went to high school in Switzerland, which is one of those places where you get bags full of live frogs from the restaurant supply house.) That is, he would take a frog out of the bag, poke it in the neck with a poker, sever its spinal cord, scramble its little froggy brains, wipe the froggy urine off his pants (frogs are understandably upset when all of this happens, and they tend to pee as a result), and then we would "reduce" the frog to his froggy parts. We'd slap him (or her—it's not all that easy to tell with dead frogs) on a table, pin his/her legs back, and slice him/her (we need a new pronoun in English; let's go for *it*) open to reveal its parts.

Then we'd take some froggy parts out, like froggy hearts and lungs and spleens (Do frogs have spleens? I told you I barely squeaked through

19

this course) to understand what makes the frog work, or rather, what made it used to work, since it has stopped working at this point. Then we reduced the heart to heart parts, and then we reduced the heart parts to heart part parts, and so on. Heart, blood, blood cells, molecules, atoms, then protons and electrons, and so on down to the tiniest of particles. (If I left out a stage, write me nasty letters, and then look to see if this is a book about biology. I didn't think so.) So, anyway, we'd reduced the frog—that's reductionism, and that's how we understand things better. We could do the same thing to you. (Do I have a volunteer? Anyone?), reducing you to your you parts to see how you work, hoping that we could make you start working again at some point, which rarely happens.

When Newton created or discovered or did whatever he did to come up with calculus, he also started (shortly after getting hit on the head with an apple, or was that Benjamin Franklin? No, that was lightning. Never mind.) to discover (not create) the rules that made nature and the universe tick along like a clock. He discovered the laws of planetary motion, laws of gravity, light, all kinds of incredibly precise descriptions of the way nature worked.

In short, Newton inadvertently described a universe that he didn't really believe in, that is, a mechanistic universe that didn't need a god to start it up or keep it running. Newton kinda liked God, as the good English Protestant he was, but somehow all the rules he discovered just sorta eliminated any need for that God, or any god at all, at least in the Newtonian universe the way it came to be understood.

It was a universe full of wonderful rules that made it act just like a machine, predictable, chugging along through all infinity without the need of any god to kick-start the motor. In fact, the rules were so beautiful that not only did the universe run like a machine, it was predictable way off into the future.

Deterministic

In the same century that Newton died (cue violins, send flowers), Pierre Simon Laplace came up with the deterministic universe. Determinism says that everything is completely predetermined by the mechanically perfect laws of a Newtonian universe. Not only can we look backward into time to see why things happened, we also theoretically should be able to look forward into time to predict everything that will happen. As Laplace said,

"We may regard the present state of the universe as the effect of its past and the cause of its future. An intellect which at any given moment knew all of the forces that animate nature and the mutual positions of the beings that compose it, if this intellect were vast enough to submit the data to analysis, could condense into a single formula the movement of the greatest bodies of the universe and that of the lightest atom; for such an intellect nothing could be uncertain and the future just like the past would be present before its eyes." [2]

We can paraphrase this into something much shorter and sweeter: if you know the location and speed of every particle, you could predict the rest of history. Everything is "cause and effect." Everything has a predictable cause, and every cause has a predictable effect, which then becomes another cause, which then becomes another effect. Since the universe is infinitely old, there is no first cause, just an endless series of causes and effects stretching back forever.

A deterministic universe has two significant characteristics. First, there is no such thing as free will. Every decision you think you make out of your own head and desires is actually just a product of many small mechanical influences the universe dictates to you. You even think you make free-will decisions because the universe wants you to think so, because you live in a time and place where having free will is very important culturally, socially, and even spiritually.

And second, a deterministic universe does not need a god. Nobody is free, everything is predetermined by cause and effect, all the way back into the infinity of time, all the way into the eternity of the future, determined by the precise laws that rule a mechanical universe, and you can figure it all out just by taking things apart to the smallest of parts and predicting what they are going to do.

The Implications

Infinite. Deterministic. Reductionistic. Mechanistic. That's the universe of Isaac Newton. That's the basis of science as most of us studied it, even today when machines aren't all that cool anymore, they're just irritating and expensive when they don't work, even today when we think that free will is about the most important thing since, well, free will, and we're pretty darn sure we've got lots of it.

To sum up: What are the implications of this IDRM universe?

First, if the universe is forever *infinite* in time and space, and given an infinite amount of time and space, anything and everything can and will happen by random chance. Eventually every possible effect will find its cause and anything that can happen, will happen.

Second, if the universe is *deterministic:* we have no free will; every decision is made for us by the random winds of math and physics, plus chemistry, biology, psychology, sociology, anthropology, and theology. And sometimes geology, if you get caught in an earthquake or a volcanic eruption. And I suppose by meteorology, you know, like when the road is wet and you slide into a tree and face-plant into an airbag. And by probably some other –ologies that I can't think of right now.

Third, if the universe is *reductionistic,* we are no more than the sum of our parts, and we can be understood just like everything else, from the universe itself to the tiniest of subatomic particles, by taking us apart to our smallest parts. OK, it's messy, and we are harder to put back together than a toaster, but still. And that's not just biologically—that is, understanding what makes our hearts beat and our spleens spleen, or whatever it is that spleens do—but also why we make the decisions we make, why we think the thoughts we think, why we react the way we react. Everything can be understood as just part of the machine that is you.

And fourth, if the *mechanical* laws of physics, underpinned by rigid mathematics, control the motion of each one of the particles we are made of, then every particle's motion is completely predictable as it bounces off uncounted numbers of other particles'; the motion of those particles is completely predetermined by previous particle collisions; we are nothing more than the sum of all those particles and all that motion; our lives, the lives every human who has lived and will ever live, and the very path that the universe takes through time is predetermined so that in the infinite vastness of time and space, random events eventually will produce every possible combination of life and existence, including the one we now inhabit.

In the Newtonian IDRM universe, we are a big accident, our existence a tiny blip in the vast, eternal history of the universe. We are nothing more than colliding particles on a cosmic pool table. There is no need of any god to produce any of this. It just . . . is.

Or so we thought, until Albert Einstein came along.

Albert
Einstein
as lead
singer for
the rock
band
KISS

23

Fill in the blanks:

1) A mechanistic universe is _____

2) A reductionistic universe has two qualities. They are _____

_____

and _____

3) A deterministic universe means _____

_____

4) The implications of an IDRM universe are:

a. _____

b. _____

c. _____

d. _____

5) True or False: physicists are much cooler than geologists because physicists play in rock bands and geologists play in bands of rock.

2

Breakthrough: The Special Theory of Relativity

Normally when people start off writing about Albert Einstein, they talk about how he struggled in school and got bad grades in math. Let's pretend I just didn't do that, so we'll start off by saying that Albert was a smart guy with hair issues.

Hair issues

Albert Einstein   Not Albert Einstein

Albert sat on a stool working in the Swiss patent office in Bern without all that much to do. Since all the machines had been invented in the 1600s while Newton was around, it was pretty dull for Albert in Bern in 1902, working as a "technical expert third class." When he discovered the special theory of relativity in 1905, one of the greatest scientific discoveries in all of human history, his bosses in the patent office were understandably thrilled, so they promoted him to "technical expert second class." That's just the kind of swell people they were.

Anyway, we're getting ahead of the story. Albert sat on his stool (metaphorically speaking) and, without all that much to do, did a lot of thinking and scribbling. He came up with some of the most amazing theories every theorized, theories about space, time, gravity, and matter so amazing that it took decades to begin to prove that they were true, since the machines we needed to prove them hadn't been invented yet. Seems like if Albert had been a little more diligent at his real job in the patent office, well, I'm just sayin'. . .

Love and Space Travel

OK, once upon a time there were two young people who met and fell in love. Their names were, well, I don't really care, so let's just make some up. Bertha Thicklobes and Eugene Bottlewattle, with apologies for all the Berthas and Eugenes of the world for whom it will seem like I am making fun of your names. I make no apologies to the Thicklobes or Bottlewattle families, who should have changed their last names a long time ago.

Here are pictures of Eugene and Bertha as college students:

As you can tell, the cream of the crop. So they fall in love and decide to get married, as people will do. They get married when they are twenty years old, by the way. That'll be important in a minute.

They go happily off on their honeymoon, and when they return, they discover very quickly that it worked, and it's Baby Time. How exciting.

How is this relevant? is what you are saying right about now. You'll find out.

So they head in for the ultrasound and discover not one, not two, but three little heartbeats, that is, not one baby with three hearts, but three babies.

Aren't they precious? If only. You might check out Eugene's and Bertha's gene pool to get a hint of the future.

In the midst of all this excitement, Bertha (who in addition to her great beauty is quite the science ~~geek~~ whiz) has applied for and been accepted by a new space program that is going to send her and her fellow geek-o-nauts out into deep space. So while she is waiting for the babies to join the human race (as if), she is in training for the space flight.

Bertha and her astronaut buddies wearing really old spacesuits.

It just so happens that the day the babies are born is blast-off day, so Bertha hops off the gurney and heads for the spaceship, and off she goes, out into deep space. She's gone a long time, and she's going very fast.

Meanwhile, back on Earth, the little ones are growing fast.

And Eugene is not having a field day, if you know what I mean.

After a long time, the spaceship comes back.

Bertha is extremely excited about seeing her family again, of course, since it has been so long. In fact, she's been gone for 30 years, she's (do the math) 50 years old, and she is eager to see how beautifully her children have grown up. Bertha clearly knows a lot more about flying in space than she does about genetics.

It must be said that being in space for a long time is really, really good for you. You know, good nutrition, low gravity, that kind of thing, and in Bertha's case, as you can see from the picture, being taken over by an alien life form has helped:

(Bertha getting off the ship)[4]

28

But Bertha gets a shock when she sees her kids come tottering over to meet her, looking more like the Sigourney-Weaver-alien than the My-Mother-is-an-Alien alien.

5

And oh my gosh, here comes Eugene.

He doesn't look so good.

In fact, he looks so bad that Bertha does a cruel thing. She asks to see his ID. Cold. Brutal. But effective, because Bertha finds out that it is, in fact, Eugene, but that he's *80* years old. And when she checks on the kids, they are all *60* years old.

But she's only been gone 30 years! She and Eugene were both 20 when she left, and she's now 50. Suddenly she's got an 80-year-old wrinkled-up, pruny husband who looks like he spent most of his life in the bathtub, and she's got kids who are older than she is!

This makes no sense. Unless you know a little bit about Albert Einstein.

Doing the Math

Bertha and her team on the spaceship were gone for 30 years, and they were going very, very fast. They were traveling at 90% of the speed of

29

light. (We call the speed of light "SoL" just to be short, and a little rude.) Let's say that in round numbers, the SoL is about 300,000 kilometers per second (kps; uh, oh, we've got two abbreviations—I always get confused when they start to do that). So 90% of the SoL is about 270,000 kps, which is (here comes one of those scientific, technical terms) very, very fast.

When you are going that fast, then things get Einsteinian. Albert discovered (while sitting around thinking on a stool in the patent office) that (this is a biggy) *time changes when you get close to the speed of light*. It slows down relative to your starting point.

Here's how it works (not *why* it works, but *how* it works).

At 270,000 kps,

5 years = 10 years

So since Bertha was gone for 30 years, 60 years passed on Earth. She got 30 years older, we all got 60 years older. It doesn't make any sense, it's weird, but it's true.

When you get a lot closer to the SoL, it gets worse.

At 299,999.99999996 kps,

10 minutes = 40 years

At 299,999.99999999998 kps

20 seconds = 80 years

Now, I know what you're thinking. You're thinking, this guy's watched one too many *Star Trek* episodes. That may be true, but you can prove that Time Dilation (that's what it's called) is true, but Einstein couldn't, because he turned down the patents for B-52 bombers and atomic clocks, both of which you need to show that this is true. Actually, some of that was a lie, but not the parts about needing B-52 bombers and atomic clocks.

What scientists eventually did was take two atomic clocks (which are even more accurate than Swiss clocks, which is why the Swiss turned down the patents—they're very touchy about that kind of thing), put one on a B-52 and one on the ground someplace safe, and flew the B-52 around in circles for a month or two, being refueled in midair and probably emptying the toilets and dropping off some Snickers. When they landed and compared the clocks' times, they found that they were off, and by the predicted amount. Time slowed down on the plane relative to the time on the ground.

Weird, but not as weird as it's gonna get.

5 years = 10 years

10 minutes = 40 years

20 seconds = 80 years

Look at the numbers above for a minute. You can see that the numbers on the left are getting smaller and smaller—5 years, 10 minutes, 20 seconds. As the speed increases, those numbers continue to get smaller until they go, at the speed of light itself, all the way to zero.

Plus, the numbers on the right side get bigger and bigger, so that at the speed of light, the right side goes all the way to infinity.

That means that at the speed of light, *no time at all equals all the time there is, all at the same time.* **Zero = infinity**. Any particle of light experiences in a single instant all of the time there has been, is, and will be, or at least until someone turns out the light. Using the Hubble Space Telescope, scientists have taken a picture of light that has been traveling for over 13 billion years. Each particle of light in that beam of light experienced all 13 billion years in a single instant. A photon (particle of light) that left the sun 8 minutes ago and reaches your eye experienced it as an instantaneous moment, sun to eyeball instantly, even though you know that it took about 8 minutes.

It might give you a headache to know that a 300,000-kilometer-long beam of light enters your brain every second. Ouch. And then your eyes and your brain get together to process 10 million bits of information every second. Ouch again.

Wait, it gets worse.

Time and Space Get Hitched

Albert also figured out that Space and Time are not separate things, but are connected into one sort of fabric, kind of like a piece of cloth, all woven together. Now, that really doesn't make any sense, 'cause the way you and I think about space is that space is, well, empty, you know, nothing there. That's why we call it space. It's full of . . . space.

It turns out that space isn't really nothing. It's . . . something. Even if it has absolutely nothing in it, no stars or planets, no space dust, no particles, nothing but vacuum, it's not really nothing. It's something, just like time is something. We call it Space-Time. (We were gonna shorten it to Spam, but the copyright was already taken.) And both of these somethings, space-time, are connected in some weird way, so that when you get close to the speed of light, not only does time change, but space changes, too. Just as time slows down, space gets squished by about the same factor.

So at 90% of the speed of light, when 10 years on the spaceship equals 20 years back here on Earth, everyone on the ship is squished by 50% in the direction of acceleration, and not only everyone, but everything on the spaceship, and not only everything, but the spaceship itself, and not only that, but the very fabric of space-time itself will be compressed by 50%.

So if we were to look at Bertha on a TV monitor, she would look all squished and flattened. But it turns out that if Bertha looks at herself, she looks just like herself. That is, she's not squished at all. And if she were to measure time, she wouldn't notice that time had slowed down. As far as she is concerned, life on the spaceship is going along at the same speed it always did, and she looks just like she always did back on Earth.

Bertha unsquished   **Bertha squished**

That's Relativity

That's why it's called the Theory of Relativity—relative to us back on Earth, she looks squished and she's living at half the speed we are back home. That's why she was so surprised Eugene and the kids were so old—she hadn't read the little card on the back of the seat in front of her that warned her about Space-Time dilation, plus where the emergency exits are and how to use her seat cushion as a flotation device in case of an emergency landing in water.

What's more, weight also changes; well, not weight exactly, but mass. Mass is not what you weigh, but how much energy it takes to move you. It takes the same energy to move you on the moon as on Earth— even though you weigh less on the moon—and it takes the same energy to move you in space, even though you are weightless. But if

33

you are already moving along pretty quickly, say at 90% of the speed of light, it takes twice as much energy to accelerate you as it does on Earth. And the faster you go, the more energy it takes.

So, for you to reach the speed of light, you would have to have infinite energy to accelerate infinite mass; you'd need a lot of time to do this, but you wouldn't have any time at all; and you would be flattened to two dimensions, except of course to you yourself everything would look normal.

And that's the Special Theory of Relativity. It's about Time.

Questions:

1) What happens to time and space relative to your starting point if you are travelling at 90% of the speed of light?

2) If you could travel at the speed of light, what time would you experience?

3) Describe what happens to the space you are occupying from your perspective, and from the perspective of earth observers.

3

From Gravity to Big Bang: The General Theory of Relativity

For Albert, the Special Theory of Relativity was the easy part. The General Theory took him about ten years to figure out—along the way, he had to go learn a bunch of math that he forgot to learn in school. (Pay attention in math class! It could save you ten years of hard work.) The Special Theory is all about time; the General Theory is all about gravity.

Gravity is so complicated and mysterious, it took Albert a long time to figure it out, and when he did, he found the solution to the problem by looking for the beautiful answer, the answer that was just so cool it took his breath away. He established what has become almost a rule in physics: expect the beautiful, simple answer to be the right one. We'll talk more about that at the end of the book.

Gravity

Here's what he discovered about gravity: it has the same effect on time and space that high speeds do. That is, gravity bends and warps space-time. That means that anything that has mass (the universe itself, galaxies, planetary systems, stars, planets, elephants, rhinos, sumo wrestlers, you and me, and even tiny particles) bends space-time around itself.

If you think of planets and stars and things floating in space-time kinda like bowling balls sitting on the surface of a trampoline, that'll help. Just like a bowling ball warps a trampoline's surface, so do stars and things warp space-time. A marble bends the trampoline a little. So the moon bends space-time a little. A bowling ball bends the trampoline a lot, and so do the sun, all the stars and galaxies, and so on.

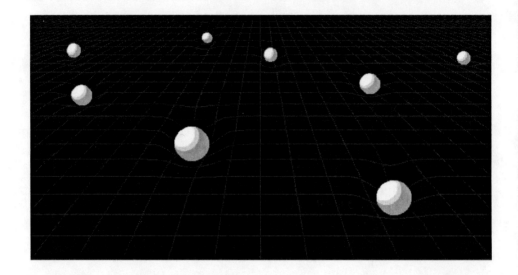

One of the cool things this means is that time passes differently the closer you are to the center of gravity, that is, to the center of the Earth. It passes more slowly in the basement of a tall building than it does in the attic. You can check this by taking your atomic clocks again, putting one on the roof and one in the basement, and just lettin' 'em sit for awhile. Before long, they'll be off.

Pretty quickly, folks started to realize that this was more than just interesting—it meant something that turned out to be disturbing. In fact, it turned the science world upside-down and inside-out.

Stars and Black Holes

Here's what happens: Stars kind of live on the edge of destruction all the time. Their immense mass translates into immense gravity, so all of the matter in the star rolls downhill into its own gravitational well, which you might be able to picture by thinking back to the balls on the trampoline. If there is only so much matter, then you just get a big rock, like a planet. If there's quite a bit more matter rolling down the

36

hill, then the pressure of all those particles on each other starts a nuclear reaction in the heart of what is now a burning star. The nuclear fusion reaction keeps the star from collapsing all the way, and the gravity keeps the nuclear reaction going.

All of that is what makes stars burn. Eventually, when a really large star burns off all of its fuel, it can no longer support its own weight, and it collapses the way it's been trying to do for a jillion years and causes a huge explosion. We'll talk about that later in this chapter.

When this happens, a large amount of stuff is blown off into space, but a lot remains behind, trapped in that gravitational well, and there can be so much gravity that, well, all of space-time in that small area of space gets pulled in with it.

You might get the idea, though, that space-time gets sucked down some sort of cosmic drain, and that's kind of true, but it's a four-dimensional drain. So it goes down into four dimensions, not down the sewer pipe, and it shrinks away to absolutely nothing at all, down past pinpoint size into no size whatsoever. Nothing.

And that's when you get a black hole. A black hole is the place where there is no place. Space and time cease to exist in a black hole. Read that again. The clock stops ticking. Time is no longer passing. It goes to that zero-equals-infinity place. And the space just goes away altogether.

If Bertha is sailing off on the flat spaces of space, then she's got no worries. If she and the rest of the crew decide to do the dumb thing and go see what's in the black hole and they cross a certain point, then they're in deep doodoo. That's called the Event Horizon, and it's the point of no return.

I read this article once called "Black Holes Don't Suck" that said you don't get sucked into a black hole, you mostly have to make a dumb decision: "Hey, let's go see what's in the black hole! We'll be famous! We

37

can send pictures back with our cell phones!" Not a good idea, but it might be kind of cool.

From our point of view, watching on our super-long-range video camera, as soon as they hit the event horizon, Bertha and her crew would stick there and stay forever. They'd never move again, frozen in time. From their perspective, they would see the rest of the history of the universe. Then they would get squished into nothingness, we think, maybe. We're not completely sure. Any volunteers? Just think . . . like everything else that happens to float into the region of a black hole, you'll roll downhill into its gravitational well and then see forever, in the place where there is no place, in a time when there is no time. Maybe.

 One possibility is found in an article called *Voyage into the Vortex: Survival Tips for Black Hole Travelers* from www.space.com: "The singularity inside a black hole is destructive, and any physical object which approaches it is necessarily and unavoidably pulverized by it. Specifically, any such object is infinitely stretched in one direction and infinitely squeezed in two other directions. That is sometimes called 'spaghettification.'" So our volunteer can look forward to being spaghettified.

Suddenly, Einstein's universe held out the possibility of catastrophic gravitational collapse into a black hole, with space-time itself being drawn into the hole. That was disturbing enough, and then . . .

The Universe Expands

A guy named Edwin Hubble started fiddling around with telescopes and made a discovery that shook the universe. He found that the universe is expanding away from us in all directions. All of the galaxies are moving apart. In fact, it's not just that they are moving apart, but that the fabric of space-time itself is stretching. The universe itself is getting bigger.

Hubble used something called the Doppler Effect to make this unbelievable find. The Doppler Effect is what happens when you see a train coming toward you and it blows its horn. As it gets closer, the sound has a higher pitch, and when you jump out of the way just in time and it passes you, the pitch drops lower.

What happens is that the speed of the train is added to the speed of sound, so that the sound wave is compressed into your ear. When it passes you, the speed of sound has the speed of the train subtracted from it (since it's going away now), so the sound wave is now stretched and sounds lower.

The same thing happens with light. When a train comes toward you and shines its light, the speed of the train compresses the light wave into your eyes so that it looks bluer—the light wave is shorter, so it shifts toward the blue end of the light spectrum. We call that the blue shift. (I saw a red bumper sticker once that said, "If you can read this and it looks blue, you're driving too fast." Subtle, but clever.)

As the light passes you, the speed of the train then stretches the light wave so that it shifts toward the red end of the spectrum. Thus, it's called the red shift. Hubble found that all the light from all the galaxies is red-shifted evenly in all directions. So all of the galaxies are moving away from us.

This means one of two things. Either we are the center of the universe—and I don't want to burst your own personal ego bubble, but that's not the case—or the fabric of space-time itself is stretching.

You can understand this by going down to the balloon store, like I did, and buying a balloon with stars painted all over the surface. If you can't afford a balloon with stars, then take a needle and poke little holes all over a cheaper balloon. Wait, that won't work. What was I thinking? Draw little dots all over the balloon. That'll work better. Then blow the balloon up.

As you blow it up, you can see that all the dots are getting farther and farther apart because the balloon is stretching. Now you have to do a

39

hard thing. You have to think of the four dimensions of the universe as the surface of the balloon—not the inside, just the outside. That's the universe, and that's how it's stretching. Space-time is stretching—space and time both—whatever that means.

The Singularity and Big Bang

When Einstein and the boys down at the patent office got ahold of this, things started hopping and popping. OK, I have to tell you that he'd been gone from the patent office for a long while by the time Hubble made his discovery, but we'll just pretend.

Scientists realized that if the universe was like a big balloon continuously expanding, then just like a balloon before you start blowing it up, at some point in the past it had to be completely collapsed. When all of the matter in the universe was that close together, the General Theory of Relativity (all about gravity) would make something frightening happen: *All of space and all of time would collapse into a black hole.*

That means that at some point in the past, all of space and time was in the same place.

Every black hole has a singularity—that's the place and time where there is no place and time, the place where space and time cease to exist. Hubble's expanding universe put together with Einstein's gravity meant that at some point in the past, all of space and time was in . . . the Singularity.

The Singularity was the place where there was no place, the time when there was no time.

There was no space.

No time.

No vacuum.

No emptiness.

No universe.

There was . . . nothing at all.

The Singularity was a single, dimensionless point of pure energy potential—no energy, no matter, no mass, nothing there but potential, whatever that means.

And then, whooooosh. Big Bang. When the whole universe came into being in a single tiny fraction of an instant. And the clock started ticking.

Here's what you may be thinking, 'cause it was what I was thinking. You're thinking that there was this little bit of dirt or something hanging around in the vast emptiness of the universe that suddenly just, like, exploded and then there was, I don't know, planets and stars and, sooner or later, duck-billed platypuses and Wal-Marts.

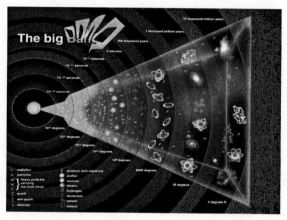

Wrong, wrong, wrong.

The little bit of dirt or something wasn't dirt or anything. It was just energy potential. And it was far tinier than a piece of dirt. It actually wasn't even there, not in the sense that we understand things to be *there*, especially since there was no there there for it to be there in. So there.

The vast emptiness of the universe didn't exist yet. Its potential was in the Singularity, waiting to be released or whatever. The clock was not ticking. Time had not yet begun. Whatever that means.

And then, whooooosh. In a bit of time almost too small to measure, a *universe*.

OK, it was empty. All we really had at that point was an expanding universe, light, heat, then darkness (after the light faded), rules (the laws of physics), and because of those rules, little tiny particles called quarks. Lots of them. We'll get to them in the next chapter.

Here's an excerpt from a 1988 article by Gregg Eastbrook called "Science sees the light" in *The New Republic*:

Suppose you accept the Big Bang theory of the origin of the universe. Here's what you believe:

You believe that, once upon a time, all the potential of the cosmos—all the potential for a firmament of 60 billion [now well over 140 billion - af] galaxies at last count—was packed into a point smaller than a proton. You believe that within this plenum of the incipient cosmos was neither hypercompressed matter nor superdense energy nor any tangible substance. [That is, there was nothing there. - af]

Next, you believe that, when the Big Bang sounded, the universe expanded from a pinpoint to cosmological size in far less than one second—space itself hurtling outward in a torrent of pure physics, the bow wave of the new cosmos moving at trillions of times the speed of light. You believe that this process unleashed such powerful distortions that, for an instant, the hatchling universe was curved to a surreal degree. Extreme curvature caused normally rare "virtual particles" [quarks - af] to materialize from the quantum netherworld in cornucopian numbers, the stuff of existence being "created virtually out of nothing," as Scientific American once phrased it. . . .

For sheer extravagant implausibility, nothing in theology or metaphysics can hold a candle to the Bang. Surely, if this description of the cosmic genesis came from the Bible or the Koran rather than the Massachusetts Institute of Technology, it would be treated as a preposterous myth.

Something extremely grand must have called forth our firmament, and whether that something was natural or supernatural may be mere semantics [since nature didn't exist yet! -- af]. Reflecting on this, Allan Sandage, one of the world's foremost astronomers, recently proposed that the Big Bang is best understood as "a miracle" triggered by some kind of transcendent power. The Nobel Prize-winning physicist Charles Townes, chief inventor of the laser, suggests that "to think that science already knows enough to be certain there are no mystical forces is illogical."[6]

God and Big Bang

You can find a lot of highly respected scientists who disagree with the idea that Big Bang offers evidence of God's existence, and a lot more like those above who say that the evidence for God is there to be found. Einstein himself didn't like Big Bang (or Black Holes or Singularities), and neither did most of the other major scientists on the planet. The late Sir Fred Hoyle, one of the most significant physicists of the twentieth century, gave Big Bang its name by saying something like, "So, it's like, what, a big bang or something?" He thought it was a stupid idea to consider that the universe may have had a starting point in space and time. In fact, he put it in stronger terms than that, saying "The passionate frenzy with which the Big Bang cosmology is clutched to the corporate scientific bosom evidently arises from a deep-rooted attachment to the first page of Genesis, religious fundamentalism at its strongest."

Einstein put a term in his equations he called "the cosmological constant," which forced the universe to be infinite, with no starting point. After thirty years, he took it out with the statement that it had been his "biggest blunder" to include it. The universe had a starting point. Space and time both began about 13.7 billion years ago. The universe is not infinite in age, and although it may be infinite in size now (and we'll never know if it is), it didn't use to have any size at all.

Where is the evidence for Big Bang? It is not overstating the case to say that the only evidence that exists supports Big Bang as the origin of space and time. There is no physical evidence for any other scientific explanation. There are no other options.

Please note, and this is important: *We're not looking for proof of God's existence in our journey here.* We are just looking to see if reasonable, intelligent, educated people can find evidence for the existence of God, and clearly in the case of Big Bang, it is a reasonable conclusion to reach. It's not proof, but it's reasonable evidence.

When Science Was Not

This is an interesting place for science to find itself, because science itself didn't exist at the Singularity. The laws of physics themselves came into being in the primal moments of Big Bang. One of the appropriate criticisms of religious peoples comes from what is called a God-of-the-gaps explanation; that is, every time humankind finds something they can't explain using science, they call it God. So we used to worship the sun, fire, earthquakes, volcanoes, and so on because we didn't know why they worked. So we called it God.

Eventually we would find some science to explain what was going on. Now physicists are saying that we may have a consistent scientific explanation for the existence of the universe and all that is within it. Thus, some may say, we have no need of God.

If we stand at the Singularity—the place where there was no place, the time when there was no time—we find ourselves in a different kettle of fish (except without a kettle or any fish). Science doesn't exist yet, so we can't look to science to explain it. By definition, our explanation would be supernatural, since we have no nature yet. We are forced into a place where we have to explain not only the origins of the universe, but the origins of science.

Where did the rules come from? And why do they work so well?

Actually, there may have been one little bit of science that was working at the Singularity. That comes in the next chapter.

In Review

Before we go there, let's have our first review. Newton's saw the universe as infinite, deterministic, reductionistic, and mechanistic. Such a universe, it was thought, does not need any kind of creator—it just is and always has been, an infinite machine working predictably forever.

Now, Einstein has shown us (to his own dismay) that the universe had a starting point, and what that means is unbelievable: space and time actually have a place and time when they both started. Before Big Bang there was no Before—no time, no space, no laws of physics, no matter, no energy. Then in a single instant, a universe and all of the energy it would need to produce everything that now exists, and all of the laws it needed to do so.

A finite universe, at least in one direction. Infinity is *history*.

Questions:

1) What impact do gravitational bodies have on space and time? What constitutes a gravitational body?

2) What does that have to do with the arrival of a Singularity?

3) So what impact does a Singularity have on space and time?

4) How does the Doppler Effect force Big Bang on science and nature?

5) What was the state of space and time at The Singularity?

4

The Quantum Netherworld: Tiny Things Playing Tricks

A few years ago there was a TV show called *Quantum Leap*. Quantum Physics has nothing to do with that. Mostly.

The word *quantum* is a Latin word that means "discreet." For our purposes, it's going to mean little, bitty, tiny things—things like molecules, atoms, subatomic particles (protons, neutrons, electrons), and sub-subatomic particles (quarks, hadrons, muons, leptons, gluons, morons, Klingons—all of that stuff). Mainly we're talking about photons (particles of light) and maybe the occasional electron. In the last few chapters, we were talking about the biggest things, a.k.a. the universe. Now we are talking about the very smallest things, a.k.a. tiny particles.

The only difference between you and tiny particles is that they are tiny and you are not, plus you are made up of them. Since you are made up of tiny little particles that have somehow arranged themselves to look like you, Reductionism would tell us that if we want to understand better how *you* work, we should just take you apart to your smallest particles to see how *they* work. However, the world of the very tiny is very different from the world you live in. By the time we get finished, you'll think *quantum* is another word for *bizarre*.

How It All Started

Quantum Physics is also called Quantum Theory or, more popularly, Quantum Mechanics, QM for short. It all started with a simple little experiment called the Two-Slit Experiment, originally conceived by Thomas Young. It looks like this:

First, you take a flashlight or some other light-shining device and use it in a set-up like the one shown:

Figure 1

You've got the light on the left, a piece of wood with two slits in it, and another piece of wood on the other side.

If you cut the slits so that they are narrower than the wavelength of the light passing through them, you get something that in classic physics we call "interference." It's like throwing two pebbles into a smooth pond. As the ripples move toward and run into each other, they get messy. That's interference.

When you shine the light through the two slits and onto the wall, it looks like this:

Figure 2

The same way ripples in water run into and interfere with each other, light waves run into and interfere with each other. An interference pattern is the result.

Physicists Run the Test

Now, physicists—instead of going out on dates with girls, which they really just ought to do more of—were sitting around thinking one day and decided to let the photons of light go through the slit one at a time instead of all at once in a wave of photons.

What ought to happen is that the photon goes through one or the other of the two slits, or misses the slits altogether and lands on the first wall. If the photons go through the slits one at a time, they should form a pattern where there are two bands of light, like in Figure 1 above.

So the physicists started shooting the photons through the equipment, one at a time. What they got looked like this:

This is after ten photons

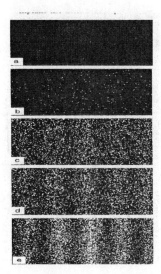

100 photons:

3000 photons:

20,000 photons:

And finally, 70,000 photons:

As you and the confused and baffled physicists can see, there are not two bands of light. In fact, the whole thing looks just like Figure 2.

That's not what it's supposed to look like. That's what you get when you let all of the photons go through at once (by turning the light on), and the waves of light interfere with each other.

49

Waves go through both slits, interfere with each other, and give us an interference pattern. Single photons don't have anybody else to interfere with. So why do you get an interference pattern? What's up with these photons? They should go through the right slit and land on the right side, or through the left slit and land on the left side. This makes no sense. I'm confused. You're confused. The physicists are confused.

This is where it starts to get weird. Here's what happened (you're not going to like it):

Each photon went through both slits at the same time. Then it kind of interferes with itself, sort of acts like a wave, and then lands like a particle.

It didn't go through one slit, turn around, come back, and go through the other. It didn't split in half so that each half went through and then they reconnected on the other side. It's just like throwing a rock toward two windows and having one rock go through both windows at the same time, without splitting in half. Both windows. At the same time.

And then it landed on the screen as one photon. Because that's what it was the whole time. One photon. Two slits. The same time.

Physicists Try Again

You might guess that the physicists weren't completely thrilled with this. So they decided, instead of giving up and going out for pizza with cute girls, to try another experiment.

They put a little electric field on one of the slits, so that when a photon went through, the field would sense it and tell the physicists what the photon was doing. So they turned on the electric field, and this is what they got:

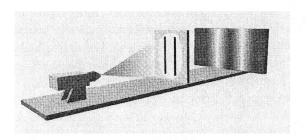

Figure 4

That is, each photon was now doing what the physicists though it would do in the first place: go through one slit or the other, but not both.

Well, *that* wasn't supposed to happen, and it was pretty aggravating. So they turned off the electric field, and the whole thing went back to an interference pattern again.

Whenever they were looking (using the electric field), they got two bands of light. Whenever they turned off the field, they got an interference pattern.

This is where it gets weirder. Here's what was happening (you're not going to like it):

The photons knew they were being looked at.

And they changed what they did depending on how they were looked at.

If you weren't looking at them trying to go through both slits at the same time, then they would go through both slits at the same time.

But if you tried to catch them doing it, then they stopped doing it.

Third Time's a Charm?

Now the physicists were really hacked off, and they started to rip their hair out, which is not a good idea for the whole dating-cute-girls thing.

51

So they said to themselves, Hey, we're physicists, we're smart, we've been to college. We're a lot smarter than photons, by golly, and now we're really getting mad.

So they set up the electric field again, only this time, they set it up between the slits and the screen, so that the photons would have already gone through both slits by the time each gets to the field. Then these clever physicists said, "We'll know what they're doing, and we'll trap them, ah *hah!* Take *that,* you stinking little photons!"

They turned on the electric field, and immediately they got this again:

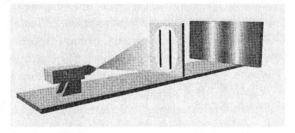

Figure 5

You could hear the physicists screaming in pain all the way to Mars. The photons once more were refusing to get caught going through both slits at the same time. But only when the physicists were looking.

This is where it gets really weird. Here's what's happening (and you're *really* not going to like this):

The photons go backward in time.

Each photon goes through both slits at the same time, gets to the electric field, says to itself, "Hey, they're trying to trap us, those devious little physicists," turns around, goes *backward in time* back through the slits, turns around and comes through either one slit or the other, but not both, and lands like a photon on the screen.

Heisenberg's Uncertainty Principle

Werner Heisenberg (didn't they name a blimp after him?) came up with what is thought to be the fundamental founding principle of QM, called the Heisenberg Uncertainty Principle. (I guess he just couldn't make up his mind what to call it.) And this is a biggie, so memorize it (or at least say it three times while standing on your head): *The speed and location of any particle cannot both be known simultaneously.*

That is,

- You can know the speed, but not the location, or . . .

- you can know the location, but not the speed, plus . . .

- a particle has neither a speed nor a location . . .

- until you look at it.

That doesn't sound very impressive, so we need to make sure we get it. I was with a friend at the International School of Bangkok once, a physics teacher, and we were talking about all of this. He remarked that what the Heisenberg Uncertainty Principle meant was that whenever you try to find out where a particle was, you mess up its speed, and if you try to find out how fast it was going, you mess up its location. Kinda like trying to find out where a fast-moving car is by running it into a wall. Messes up the car, not to mention the crash-test dummies.

That's the wrong way to understand QM. What it really says is the last two phrases: A particle doesn't have a speed or a location at all. Until you look at it.

Questions:

1) Are there any questions we can ask here that have an answer that makes sense?

2) What does it take for a particle to have either a speed or a location?

3) Can you know both the speed and location of a particle?

4) Why does this give us the potential for free will?

5

Quantum Science: Where Reality Gets Really Weird

"A particle doesn't have a speed or a location at all. Until you look at it."

One of the things that this means, right off the bat, is that *notre ami Francais* M. Pierre Simon Laplace is dead in more ways than just, you know, dead. (More on this later.) If you remember back in chapter one, Laplace is the French guy who said, "Give me the speed and location of every particle, and I'll predict the history of the universe." On that basis, he defined Determinism, which took away any free-will decision that we might ever make.

But our good *Deutscher Freund* Herr Doktor Heisenberg said, "You can't know both the speed and location of any particle." He wins. The universe is not purely deterministic. The Germans are always beating up on the French. *Tant pis. C'est la vie.*

That doesn't mean that there is no cause and effect going on in the universe. Surely if I throw you out of an airplane without a parachute, you will fall, and chances are really good that you will get squished and that you will also squish something or someone. You still make decisions on the basis of society, culture, peer pressure, parental pressure, faith (even lack of faith causes you to make decisions), the laws of nature, fashion magazines, movie stars, and accidentally stepping in something gross.

In this chapter we're going to look more closely at Quantum Physics and see how it connects to reality. Or doesn't. The real treat is, if you pay attention, you may find out how quantum physics applies in everyday, real life.

Quantum Unpredictability

Because, at the quantum level, effects happen without a cause and causes don't always cause effects, this opens the door in nature for quantum unpredictability. And since you and me and the rest of the universe are made up of quantum particles, well, who knows how that plays out? There is literally no way to know for sure why what happened in the past happened, and no way to know for sure what is going to happen in the future. When we get to Chaos Theory in the next chapter, we'll put the final nail in that coffin. At some level, you are theoretically able to make a free-will decision, if only in part. But let's not get into that right now.

There are some weird and wonderful things that go on (kinda) in the quantum world. First, here's a brief outline of the three facets of quantum unpredictability:

1. Particles Can Be in Two Places at Once.

First is the effect we've seen already: a particle being in two places at the same time. In fact, a particle can be in lots of places at the same time, close to an infinite number of places, which is to say, everywhere, all at once.

As strange as it sounds, some physicists actually did this to a particle in a lab in Boulder, Colorado, not too long ago. Here's an excerpt from an article in *USA Today*:

David Wineland and fellow researchers at the National Institute of Standards and Technology in Boulder, Colo., coaxed a beryllium atom in a vacuum to be in two places at the same time—the paradoxical "Schroedinger's cat" state. The researchers then caused the system to collapse <u>by introducing contact to the outside world</u>.

The NIST researchers said they were able to keep a beryllium ion in a Schroedinger's cat-like state for as long as 100 millionths of a second.

To do that, the beryllium atom was cooled to close to absolute zero—minus 459 degrees—and isolated from all types of radiation, radio waves and other energy sources.

The researchers then used lasers to force the atom's single electron into two states of spin, which also forced the atom to be in two places at the same time.[7]

We'll let Schroedinger's cat out of the bag—uh, box—soon, and we'll also see how this little piece of quantum-ness turns out to be potentially pretty useful. But for now, notice in the first paragraph of the article above that the atom stopped being in two places at the same time when they introduced "contact to the outside world," that is, in our terms, when someone looked at it.

2. Particles Come into Existence Out of Nowhere.

Another bit of quantum weirdness that we'll explore further later is the habit quantum particles have of just popping into existence out of nowhere. Here is a brilliant piece of causeless effect—particles just spinning into existence like a tornado out of still air.

In fact, it's worth noting that energy and matter, as Einstein put it so well in his equation $E=MC^2$, are the different forms of the same thing, kinda like (but not really) water, ice, and steam are all forms of the same thing: H_2O. $E=MC^2$ was the equation that people used to create both nuclear energy and nuclear weapons, releasing the energy that is trapped in a spinning tornado-bit, forming the illusion of solid matter. Einstein summed up the quantum world (and the real world) by saying, "Reality is merely an illusion, albeit a very persistent one."

Matter is just contained energy, like a tornado is contained air, with nothing really there to be a tornado except for energy swirling the air around (and the occasional cow). There's nothing really there to be matter except for energy spinning in a very tight orbit: the entire universe—and everything in it—is little bits of spinning energy.

How the little bits of spinning energy figure out what to do—that is, how to become matter—is what quantum physics tells us we don't know. Energy just becomes matter, randomly, out of nothingness, just because the rules of quantum physics tell them to do so. We'll talk about that more in chapter 13.

By the way, the bits of energy-that-become-matter-out-of-nothing are quarks, or at least the ones we will worry about are called quarks. A very bright physicist named Murray Gell-Mann chose the name out James Joyce's book *Finnegan's Wake*. See, physicists may not go out on dates much, but they do read things that aren't just physics. You need to know that Murray is probably going to be irritated that I called him "very bright." He's a lot smarter than that. Sorry, dude. I'm still not saying he went out on dates, though.

3. Particles Communicate Quicker than SoL

The next thing (we're not going in any order here; I'm just picking out all the things that I think are cool) that goes on in the quantum world is faster-than-light communication between particles, or at least, between parts of particles. That may not seem like a big deal to you, since they seem to do that on *Star Trek* all the time, but let me just say this the way Albert would say it: *You cannot have faster-than-light communication.* Read that sentence again forty times. Albert says that sentence is true; the theory of relativity says it's true; everything in the universe says it's true. But in the quantum world, it seems to happen, like, all the time. I'll explain why after we let Schroedinger's cat out of the box.

So, that's quantum unpredictability in a nutshell, or should I say, at faster than the speed of light.

Quantum Tunneling

The second coolest thing in the quantum world is quantum tunneling. Here's how it works. No, that's not right. We have no idea how or why it works. We just know that it . . . works.

Go down in the basement and find an electron gun. Or you can do what I did and find one on the Internet. Now take your electron gun and shoot electrons, one at a time, at the basement wall, like this:

In our little experiment here, we've measured the distance from the end of the gun to the wall so that we know exactly how far it is, and we know how fast the electrons are traveling (very, very fast), so we know exactly how long it takes them to get there.

Now, let's put a lead wall between the gun and the basement wall, like this:

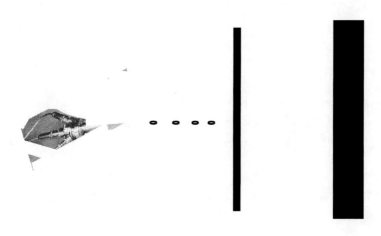

It's solid lead. Really solid. Much too solid for anything to get through, at least, anything like an electron.

So what happens is, most of the electrons bounce off. But some . . . don't.

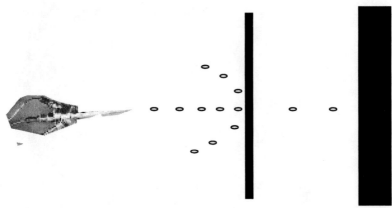

Now what you are trying to do is figure out what's going on. You might be thinking (if you are only slightly less smart than Murray) that maybe, though the electrons are really small and the lead wall is really thick and solid, every now and then an electron will get lucky and kinda bounce through and squirt out the other side.

Nope. That's not it.

Then maybe you're thinking that the electrons are going so fast that every now and then one of them knocks another one off the other side of the lead wall and off it goes.

Nope. That's not it either.

This is where it gets really weird. Here's what's happening (and you're not going to like it):

Some of the electrons just skip the wall.

They skip it. They don't go around. They don't go through. They don't go over. They just instantaneously skip from one side of the wall to the other, without traveling the distance of the wall.

You know how we know this? It's because *the electrons that skip the wall get to the basement wall faster* than when the lead wall is not there. That's because they don't travel as far when the lead wall is there. They don't travel the width of that wall. They just . . . skip it.

So, the thicker the lead wall, the quicker they get to the other wall, because they have less distance to travel.

No Reality Without Observation

Quantum unpredictability is the most bizarre thing. Quantum tunneling is the second coolest in the universe. Here's the coolest. It starts with our statement that particles have neither a speed nor a location until you look at them. That is to say: *reality is made by making an observation.* One could say (and we will, over and over again) that nothing happens without an observation. Nothing at all happens unless someone takes a good look at it.

We clearly need to define what that means, since in normal English, it means doodly, as the physicists are fond of saying (that would be another lie). So let's take a good look at Schroedinger's cat.

Schroedinger's Cat

Erwin Schroedinger, a good German quantum theorist, had something of a problem with this whole idea of there being no reality without observation, so he created a thought game to baffle and destroy his quantum enemies. Unfortunately, this particular cat landed on its feet, as it were.

Here's the game. He said, let's take a quantum particle and put it in a box. If we aren't looking at it, that is, if the box is closed, then the particle dissolves into a state of quantum uncertainty, or at least that's what all the wicked QM theorists were saying that gave Erwin a lot of hostility.

So he said, let's make this particle an alpha particle, a radioactive particle. Radioactive particles decay eventually, and so this one will decay, eventually. We don't have any way of knowing when it will decay, and if we close the box, then (the evil QM scientists said) this particle will dissolve into a quantum state of uncertainty. That means that the particle has neither decayed nor not decayed. Both possibilities coexist simultaneously.

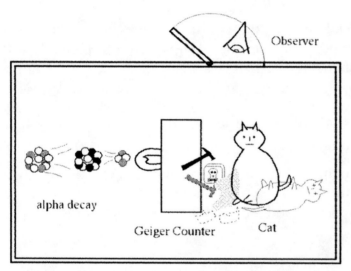

alpha decay

Geiger Counter   Cat

Now, said Herr Schroedinger, if that is true (and he thought it was craziness), then if I put some other things in the box, it will all get interesting. So, he said, let's put a Geiger counter in the box that will detect the decay. Then we'll hook up a hammer to the Geiger counter that will fall down if the counter clicks, and a bit of poison in a glass jar that will break, spilling the poison.

Then we'll put a cat in the box, and close the box so that no one can see in.

Thus, he said, if the alpha particle exists in a quantum state of uncertainty, then so does everything else. Especially, he suggested with a sneer, the cat. So, what you are telling me, he told the evil QM scientists, is that the cat exists in a quantum state of uncertainty. Which means (you're not going to like this) that the cat . . . is neither alive nor dead.

The evil QM scientists all thought about that and decided that, as crazy and weird as it sounded, it just happened to be absolutely right. That's the way the quantum world works.

If you open the box and look inside, then things get very strange.

The observer "collapses the wave function" when he or she looks inside. What that means in English is that things stop being weird and get normal, sort of. Here's what might happen:

The alpha particle, when the observer looks in the box, says to itself, Ah ha! That observer's lookin' at me. The curtain's up, the band's playing, and it's show time!

So the particle decays, or does not decay. It may not be time for it to decay yet, or it may have already decayed. It doesn't decay instantly when the observation is made. It may not decay for another 10,000 years. Or it may have decayed, say . . . last week. So, according to Schroedinger, when the observation is made and the wave function collapses, the particle decays . . . last week. This means that the Geiger counter clicked last week, and the hammer dropped last week, and the poison spilled last week. And so that means that the cat is dead, but . . . *the cat has not yet been dead for a week until you open the box.*

Reality is created when you make the observation, but you have no control over the reality that gets created. Things don't happen at the exact moment you look. They happened earlier or they will happen later, but they haven't happened at all until you look in the box.

By the way, don't expect to wrap your mind around all this—unless you're an Einstein or something, and he didn't get it either. Here's the important thing to remember. Read this over and over again until you die.

Nothing happens at the quantum level without an observation.

Nothing.

At the quantum level, everywhere in the universe and everywhen in time, nothing ever happens without an observation. Somebody has to look at it or make a measurement of some sort.

Otherwise, nothing happens. Things just sit there in a quantum state of uncertainty, waiting for an observation to be made . . . waiting, waiting, waiting.

The Copenhagen Interpretation

And it takes an intelligent observer to make the observation. The cat can't observe itself into reality. It takes (you're not going to like this, though I don't quite completely understand why) a human to make the observation. Or some sort of intelligent alien from another planet, I guess.

As you might imagine, scientists didn't like this, and they still don't. It's called the Copenhagen Interpretation of quantum theory, named after the city where a Danish QM guy named Niels Bohr lived. He was very, very good at QM.

Albert Einstein didn't like QM and he didn't like faster-than-light communication between particles, calling it "spooky action at a distance," so he used to write letters to Niels, complaining about QM. Albert would sit around for days coming up with brilliant reasoning as to why QM had to be wrong (he said he couldn't believe that "God would play dice with the universe"), and he'd mail these letters to Niels. Niels would think about them for about 45 seconds, and then write Albert, the world's all-time most-favorite intelligent genius scientist, and tell him he was wrong. And Albert was wrong, every time. The smartest guy in the universe was always wrong. What's funny is that Einstein won a Nobel Prize in QM. And he still didn't like it.

"Spooky Action": Communicating Back in Time

One of the things he disliked the most was this "spooky action at a distance," "entangled" particles communicating (which is weird all by itself) at speeds much faster than the speed of light. We've measured how fast they communicate (and when I say "we," I mean "not me, but people a lot smarter than me"), or at least we've tried to measure the speed. And the closest we've been able to get is a communication speed between particles of 10 million times the speed of light.

That's fast. And as Albert would be eager to tell you, it's not possible.

Here's what happens: You take your basic particle and split it into two halves, using very tiny scissors (that would be a lie). Then you isolate the halves of the particle so that they don't have any contact with the outside world. The cat in the hat—er, box—described above tells us that if you can't see them, the particles dissolve into quantum states of uncertainty.

If the original particle had a characteristic, which we will define in our example as being a good cook, then when you divide the particle into two parts, being a good cook dissolves into a quantum state of uncertainty. The two halves neither cook well nor badly. Cooking well and cooking badly coexist simultaneously.

Then you separate the two halves of the particles. It could be a few feet, it could be across to the other side of the universe. And then you look at one half of the particle. You make an observation.

As soon as you make the observation, that half of the particle has its wave function collapse, and it instantaneously becomes, say, a good cook. At the same moment, across the room or the galaxy or the universe, the other particle instantaneously becomes a bad cook. The first half communicates at 10 million times the speed of light (or faster) to the second half what it should do, even though you have not looked

65

at the second particle. The collapse of the wave function, that is, the observation you make on the first half, gets communicated instantly to the second half.

A real scientist named John Gribbon wrote a book *(Schroedinger's Kittens and the Search for Reality)* trying to figure this out, and he said that the only way it's possible for this to happen is not for the particles to communicate faster than light, but backward in time. The first half of the particle sends a messenger particle back in time to the moment before the two halves were divided, tells the other half what's going to happen, and then another messenger particle gets sent into the future to seal the deal. Like I know what a messenger particle is and how these little bits of nothingness are sending messages like teenagers IM'ing each other.

Here's the real answer: We don't know how they do it. They just do it. They are like the total Nike particles. They just do it.

Practical Application: Computers

OK, now you may thinking, That's pretty cool, but big wup. Does any of this turn out to be useful in any way, or relevant to the topic at hand?

Turns out that it's all useful and extremely relevant. Scientists have been talking, for example, since about 1981 about building a quantum computer. A normal computer uses binary code to work—all it is, really, is a quadrillion on-and-off switches. So you have two choices—on or off—for each switch.

In a quantum computer, where things exist in a quantum state of unreality (remember, the cat was neither alive nor dead, or, if you like, both alive and dead at the same time), you can have switches that are both on and off at the same time. That's odd enough, but when you consider that the switches are not only the size of atoms, they also are the actual atoms themselves, then things get very interesting, and tiny. That means you have atomic-sized switches that each have twice as many choices as normal computer switches, so you can double the amount of information being processed every time you add an atom.

Here's a bit of an article from *Time* magazine, February 24, 2003:

Researchers in the Netherlands and Japan reported in the journal Science last week that they had caused an electrical current in a super-conducting ring to flow simultaneously clockwise (representing 1) and counterclockwise (0). The result was a "qubit," a quantum representation of both the digits of binary arithmetic.

In other labs, qubits have been devised from single atoms. Whatever is used as the quantum abacus beads, the result is an exponential explosion in computing power. By the time you get to just 14 atoms, a speck still far too tiny to see, you can do more calculations in tandem (16,384) than the fastest supercomputer in the U.S.[8]

Fourteen atoms are not very big as a group, and as we'll see, they can be contained in a tiny drop of liquid: *"Scientists at Los Alamos National Laboratory announced the development of a 7-qubit quantum computer within a single drop of liquid, a simple fluid consisting of molecules made up of six hydrogen and four carbon atoms."[9]*

And this, from the *Scientific American* website:

Factoring a number with 400 digits—a numerical feat needed to break some security codes—would take even the fastest supercomputer in existence billions of years. But a newly conceived type of computer, one that exploits quantum-mechanical interactions, might complete the task in a year or so, thereby defeating many of the most sophisticated encryption schemes in use. Sensitive data are safe for the time being, because no one has been able to build a practical quantum computer. But researchers have now demonstrated the feasibility of this approach. Such a computer would look nothing like the machine that sits on your desk; surprisingly, it might resemble the cup of coffee at its side.[10]

I don't drink coffee, so I'm going to hold off buying a quantum computer until they can make it out of Dr Pepper, or maybe chocolate, even though that's not really a liquid.

Another important, interesting, and incomprehensible (to me) factor in quantum computing is that the "spooky action at a distance" that bothered Einstein so much is one of the critical aspects that makes quantum computing possible. I'm just going to put the paragraph from the *Scientific American* website right here so you can read it (if you dare):

Another property of qubits is even more bizarre—and useful. Imagine a physical process that emits two photons (packets of light), one to the left and the other to the right, with the two photons having opposite orientations (polarizations) for their oscillating electrical fields. Until detected, the polarization of each of the photons is indeterminate. As noted by Albert Einstein and others early in the century, at the instant a person measures the polarization for one photon, the state of the other polarization becomes immediately fixed—no matter how far away it is. Such instantaneous action at a distance is curious indeed. This phenomenon allows quantum systems to develop a spooky connection, a so-called entanglement that effectively serves to wire together the qubits in a quantum computer.[11]

Now I understand about spooky action at a distance, which is to say, no, I don't, but at least I know that no one else does either. But how this helps quantum computers to function—I'm just going to assume nobody else gets it either.

A Little Bit More Quantum Theory

Something I should have mentioned earlier is how much quantum particles don't want you to see them, that is, to know both how fast they are going and where they are at the same time (remember Herr Doktor Heisenberg?). They really don't want you to know. They'll do amazing acts of legerdemain (I think that means "magic"—every now and then I have to use a big word to impress any real scientists that might be reading this) to avoid being caught with both a speed and a location.

For example, there was a very bright Indian guy named Satyendra Nath Bose back in the 1920s who came up with a very cool (you can't imagine how cool) idea that he sent off to Albert Einstein because Bose was Indian and nobody would listen to him, even though he was almost as smart as Murray. Albert thought Bose's idea was amazing, so they wrote a paper together about something called a Bose-Einstein Condensate, or BEC. (I was always impressed that Einstein, the most famous scientist in the universe, would put Bose's name first.)

The idea was that if you cooled some atoms down to where they were very cold, QM would kick in and strange things would begin to happen. When the atoms got very, very cold, they would stop moving (no speed) and stop moving (a location). So anyone watching would be able to know both how fast they were going (zero) and where they were (right there), and quantum mechanics would cease to exist and I wouldn't have had to write this chapter.

So the atoms would have a panic attack (they just hate having anyone know that much about them) and *form a new type of matter altogether*, which Bose and Albert called a BEC. For about seventy years, it was just a good idea that no one had the technology to prove.

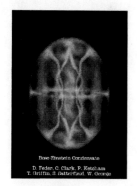

Bose-Einstein Condensate
D. Feder, C. Clark, P. Ketcham
T. Griffin, S. Satterfield, W. George

BEC

In 1995, though, a couple of scientists decided to give it a shot. Eric Cornell and Carl Wieman in Boulder, Colorado, cooled some rubidium atoms so cold that it was the coldest place in the universe—less than 170 billionths of a degree above absolute zero.

When they did this, the atoms had their little panic attack—*Oh, no, everyone is going to know where we are and how fast we are going!*—and merged instantly into a super-atom where all the atoms became exactly identical to one another. That meant that even though you knew how fast they were going (zero), you couldn't know where any of them were.

They are really serious about this speed and location thing, so serious that they formed a new type of matter that doesn't exist anywhere else in the universe, because it's not cold enough anywhere else in the universe for the BEC to exist. (Probably I should mention that the heat left over in the entire universe from Big Bang is about 2.7 degrees above absolute zero, so nothing can be naturally colder than that anywhere, except maybe on this ski lift in Vermont I rode once—you could see BECs shivering in their parkas.)

Why Does It Work?

Quantum tunneling is essential for the following physical processes to take place:

- Laser technology (that means CD/DVDs)
- Nuclear energy (both fission and fusion)
- Superconductivity
- Semiconductor technology (anything with a chip in it: cell phones, microwaves, dishwashers, trains, planes, automobiles, computers, etc.)
- Genetic engineering, cloning
- DNA and the stuff of life
- Photosynthesis and maybe some animal respiration

So we're not only talking about computers, CDs, DVDs, and nuclear energy, we're also talking about life itself being dependent on quantum tunneling to exist. The question is, Why?

I was sitting next to a Chinese physicist at a meeting in Hong Kong once and started talking to him during a break. First we talked about his two wives (you never want to have two wives alone in a box with a cat and a hammer, is what I'm thinking), then about his background. He had his PhD in solid-state physics from MIT. I kinda thought that gave him some credibility, so I asked him why quantum tunneling is necessary for silicon chip technology to work. He said, "I dunno." That was helpful.

So I went out and read about it and found out that it has to do with electrons jumping from one energy level to another in an atom and something something in the testing procedure and blah blah blah. You can go read about it, too.

Run that by me again?!?

To try to explain the whole thing a little more clearly (yeah, right), remember that little model of the atom you may have made in fifth grade, the one that looks like a little solar system with planets whizzing about? It may have looked like this one I got from the Florida Department of Health website:

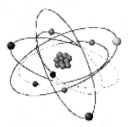

I don't want to hack anybody off at the Florida Department of Health, but that picture, and the model you made in fifth grade, are just garbage. They make you think that the electrons whizzing about have a definite location and speed.

But as it happens, if electrons were whizzing about like planets, they would quickly burn off all their energy and fall into the sun, or, in the case of the atom, the nucleus, and we would have no atoms, and it would get messy.

So what happens is, the electrons don't have a place where they are. They are smeared out in electron shells that look and act just like waves:

71

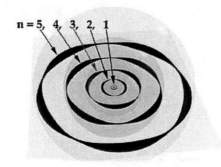

n = 5, 4, 3, 2, 1

www.wonderquest.com

Each ring in the picture above is a place where electrons hang out, but there's no place on the ring where each electron actually is, and no speed that it's actually moving at. Until you look at it. Then you can find out where it is, but not how fast it's going, or how fast it's going, but not where it is. There are places where there is a higher probability that it might be, and other places with a lower probability (which includes everywhere in the universe—every quantum particle has a low but real probability that it could be anywhere in the universe), but no place where it actually is until you look at it.

I should also tell you that one of those QM guys, maybe Niels himself, said that if you think you understand quantum physics, you just haven't thought about it enough yet. Now *that* I understand. Or not.

And this just in – New Scientist Magazine writes that if Einstein was right, then either 1) there is no such thing as free will, 2) there is no such thing as reality, or 3) information can travel at speeds faster than the speed of light.[12] Take your pick.

In Review

OK, let's put it all together then. What does QM mean for Newton's universe? Newton's universe is the IDRM universe, that is, infinite, deterministic, reductionistic, and mechanistic. Einstein's special and general theories of relativity killed off the infinite universe. What did QM do for Newton?

We've already seen that QM has shown us that the universe is not purely deterministic. Although much of what goes on in nature is cause and effect, at the smallest level, nature is not driven by cause and effect at all; things happen without a cause all the time. And since we and everything in the universe are made up of very small things, at the most basic of levels, nature is not deterministic at all. Somehow between a quantum particle and something the size of you and me, nature resolves itself into reality, but where that happens and why are some of the great unanswered questions of physics and metaphysics.

The philosophical side of reductionism is also in deep trouble. Reductionism says that in order to understand things better, we take them apart, and the smaller the parts, the better the understanding. But now we have seen, courtesy of the quantum world, that when we take things apart to their smallest parts, not only do we not understand them better, we don't understand them at all. So reductionism is in a bit of trouble. It's not dead yet; it's just wounded. We'll put it out of its misery later on.

Mechanism is also wounded, since a mechanistic universe in predictable, and the quantum universe is not. But, like reductionism, it's not dead…yet.

First, we have to Rage Against the Machine. That's a band. I don't know any of their music, but it's what Chaos Theory is all about. Kinda. (We'll get into Chaos in the next chapter, if this chapter wasn't chaotic enough for you.)

What does QM do to Einstein's universe? Einstein's universe has a Big Bang in the middle of it, or, rather, at the beginning of it. And Big Bang has the Singularity at the beginning of it.

The Singularity was a very small particle, smaller than a proton. It was a quantum particle. That means that Big Bang was a quantum event.

Remember: Nothing happens at the quantum level without an observation. Nothing.

If quantum mechanics was working at the Singularity, then something had to look at it to make the Big Bang boom. Something, or someone, some intelligent something had to make an observation, if the Copenhagen Interpretation of QM is true (and despite the prejudices of many scientists who don't like it [for obvious reasons of just pure intellectual discomfort], it is the Standard Model of Quantum Mechanics as it is taught in universities worldwide.)

Somebody may have had to take the Big Look to start Big Bang. It's not you, it's not me, and it's not Einstein. That only leaves room for the Big Guy.

It's possible, of course, that quantum mechanics was not working at the Singularity, since all of the other laws of physics came into existence with Big Bang. It's possible that it was just a random quantum fluctuation, the universe itself coming into being out of nothing because of nothing. We'll never know.

It's kind of cool to think about, though.

So pay attention. This will all come back again before we're finished.

Questions:

1) Are there any questions we can ask here that have an answer that makes sense?

2) What is matter, how does it arrive, and does it matter?

3) How long has the cat been dead before you look in the box?

4) What happens without an observation?

5) What happens with an observation? Who gets to make it happen?

6) Which is more likely to be correct – there is no such thing as reality, there is no such thing as free will, or things can travel faster than light?

7) What had to happen for Big Bang to occur? Do you buy it?

6

Chaos: Where Small Things Make a Big Difference

I need you to put this book down and go watch a DVD for a minute. No, not just any DVD. Go watch *Jurassic Park*. You don't need to watch the whole thing. Just skip to Chapter 9, "Start the Tour!" (at about 46:50) where Ian Malcolm is in the car with what's-her-name and Dr. Grant. . . . Ellie Sattler. That's it. She's the pretty blond (but, and I hasten to add, very intelligent) paleobotanist, and of course, a doctor as well.

The scene opens with Dr. Ian Malcom, who is a "chaotician" (which my spell checker tells me is really "Chaotic Ian"—now that's just a little too perfect), saying, "The tyrannosaur doesn't obey set patterns or park schedules, the essence of chaos," and the scene goes on like this:

Ellie: I'm still not clear on Chaos.

Ian: It simply deals with unpredictability in complex systems. The shorthand is the Butterfly Effect. A butterfly can flap its wings in Peking and in Central Park you get rain instead of sunshine.

Ellie gestures with her hand to show this information has gone right over her head.

Ian: Did I go too fast? I go too fast, I did a fly-by. (Points to the glass of water) Give me that glass of water. We're going to conduct an experiment. It should be still, the car's bouncing up and down, but that's OK, it's just an example.

He dips his hand into the glass of water. He takes Ellie's hand in his own.

Ian (cont'd): Now put your hand flat, like a hieroglyphic. Now let's say a drop of water falls on your hand. Which way is the drop going to roll off? Which finger or your thumb or the outside or . . . ?

Ellie: I don't . . . the thumb, let's say . . .

He flicks his fingers and a drop falls on the back of Ellie's hand.

Ian (cont'd): Ah ha, OK, OK. Now freeze your hand, freeze your hand, don't move. Now I'm going to do the same thing, start from the same, same place again. Which way is the drop going to roll off?

Ellie: Let's say, back.

Ian: It changed. Why? Because tiny variations—the orientations of the hairs on your hand, the amount of blood distending in your vessels, imperfections in the skin—

Ellie: Oh, imperfections in the skin?

Ian: Microscopic, microscopic—never repeat, and vastly affect the outcome. That's what?

Ellie: Unpredictability . . . [13]

There we go. It should all be clear to you. This chapter was easy.

Yeah, right. Still, Chaotic Ian did give us some key phrases: *unpredictability in complex systems, the Butterfly Effect, tiny variations,* and finally, just the simple word *unpredictability,* which we can amend to *Radical Unpredictability.*

First, though, let's deal with Ian and ask the profound, academically challenging and intellectually thorny question, What the heck is he trying to say? Does it have any meaning, or is he just trying to put the moves on Ellie with her boyfriend sitting right there staring out the window at what turns out to be a really large pile of dinosaur poop while his girl flirts with a slimy guy with a gold chain in his chest hairs? That's a darn good question. It'll be on the test. Plus we need to know if Dr. Grant still likes Ellie after she sticks her arm all the way to the shoulder into the really large pile of warm dinosaur poop. I'm thinkin' not.

78

Back to the point. Ian does a pretty good job demonstrating what chaos theory is all about. First he conducts a simple experiment using whatever he happened to have at the time. You couldn't do this with many of the sciences—biology, chemistry, physics, they all need a bit more specialized material and preparation.

But chaos theory applies to every form of human inquiry, from physics to physical education, from weather patterns to sewing patterns, from the stock market to the sock market—if there was a sock market. You know. A place where you buy socks. OK, that didn't work so well, but you get the idea. So Ian can just grab a glass of water and a hot babe's hand and demonstrate chaos theory just perfectly.

It was also perfect in location. Even though he complained about the car bouncing up and down, that bouncing gives us an even better demonstration of what chaos is all about. It's all about tiny variations that make it impossible to predict what's going to happen. Sometimes the universe bounces up and down.

In Ian's experiment, he drips a drip of water onto Ellie's hand, and she has to predict which way it is going to slide down her hand. Then he drips another drip, and she has to try to predict, on the basis of what happened the first time, which way it's going to slide the second time. And she misses; she gets it wrong. Why?

Tiny Variations: The Butterfly Effect

As Ian says, tiny variations in "the orientations of the hairs on your hand, the amount of blood distending in your vessels, imperfections in the skin" have changed the path of the water drop. And it's not only those tiny variations; it's the bouncing up and down of the car that has changed the position of both of their hands, even if only very slightly. It's the air moving in the car, the size and position of the water droplet on his hand, the way he holds his hand each time slightly differently before the water drops off. It's many, many tiny variations in the experiment that make it literally impossible to predict what's going to happen.

The Butterfly Effect tells us that even though most of the time those tiny variations don't have any significant impact on anything (like, who cares which way the water drip drops?), every now and then those tiny variations will change things enough that the slightly different outcome will have an impact on something else, and that something else will affect something else, and so on and so on in a cascade of events that started with something tiny, but might just result in something huge.

Jurassic Park, for example, is a great chaos movie. You start with scientists making an assumption (arrogant, as it turns out) that they are in control of their science. Blend in one hacked-off unappreciated computer geek, an unexpected storm, and the inherently unpredictable nature of nature (dinosaurs and humans together, in this case, or, if you like, carnivores and *carne*), and the whole structure comes crashing down.

The Butterfly Effect says that sometimes a butterfly will flap its wings off the coast of South America or wherever and set off a cascading series of events, initially very tiny, that can lead to a hurricane in the Atlantic. It clearly doesn't happen every time a butterfly wing flaps or we'd have nothing but hurricanes. But every now and then, it happens just like in *Jurassic Park*.

... And Bees

Let's change our butterfly to a bee, just for fun. The bee is flying around, flapping his or her little wings. Most of the bees are doing what bees do, but this bee buzzes over close to the road. It happens to be a warm day, so most of the drivers on the road have their air conditioners on with the windows closed, but the car that you are in has a broken air conditioner (maybe because some little drop of water got something wet that corroded and stopped working. Are you starting to get the idea?), so the window is open.

Now, the bee could have missed the car entirely, or bounced off the windshield and gotten squished (bad for the bee, no worries for you). But this bee got in the wrong spot at the wrong time and was dragged by the wind through the window. It could have been the back or passenger window, but it was the driver's window. The bee could have bounced off the driver's head or chest, or missed the driver altogether, but this bee went in the driver's eye.

The driver wasn't wearing sunglasses because it was a warm but cloudy day. The bee landed stinger down, and it went deep into the driver's eye.

Now you can start to fill in the blanks. But the cascading series of events could stop or continue, dissipate or not. Maybe the driver goes, "Ouch!" and that's the end of it. Maybe the driver swerves, gets control, and that's the end of it.

But maybe the driver swerves; loses control; drives off a huge cliff, lands on Air Force One that just happened to be flying below; the president is killed; the vice president takes over; he's a wicked evil sicko; he attacks Canada to punish those Canadians for letting Mad Bees come over the border; Canada goes nuclear and bombs Finland (which beat it in hockey last week, so the Canucks are all hacked off); and the world collapses into nuclear holocaust, after which the only things that manage to survive are butterflies and bees.

That's chaos, except a little silly just for our story. The Butterfly Effect works just like that, and sometimes there are huge unpredictable results from a little tiny starting point.

Radical Unpredictability and Weather

Because of chaos, the universe is radically unpredictable. Tiny variations in initial conditions make it impossible in very critical ways to predict anything. We'll look at some.

Changes in Initial Conditions

One of the first evidences of chaos found in science was when scientist Edward Lorentz was fiddling with weather patterns. Here's an explanation for what happened:

The study of chaotic systems began in the early 1960's when Edward Lorenz, a meteorologist at the Massachusetts Institute of Technology, stumbled upon a bizarre phenomenon. At the time, he was working on mathematical models for weather prediction and had developed a simple set of equations which governed the behaviour of an artificial meteorological system.

One day, he wanted to analyse a particular run of the system again. Rather than start the sequence from the beginning, he took a printout of a previous run and typed in the numbers from midway through that run. Then he went off for a cup of coffee (although his computer was the size of a fridge, it had only 16KB of memory and could only calculate at a rate of 60 multiplications a second).

What he found when he returned has generated an entire field of scientific research. The data produced by the system should have exactly reproduced the patterns of the previous run since that run started from the same state. In fact, the generated sequence diverged steadily from the previous run to the point where all resemblance had disappeared. He initially thought that one of the valves in his computer must have blown. However, this would have led to a sharp change in the behaviour of the system.

Lorenz eventually discovered that this effect was due to the short-cut he had taken: while the computer stored numbers to a precision of six decimal places (e.g., 4.389204), the printout that he took the figures from only depicted three (e.g., 4.389). He had uncovered one important property of chaotic systems: their overall behaviour is incredibly sensitive to minute changes in the initial conditions.[14]

Traditionally, scientists have assumed that initial conditions were not all that important—close enough was close enough. When you do a science experiment in the lab at school, your results will always be a little bit different as you do the experiment a number of times. You seldom get the same exact answer, but they all tend to be close to each other. Your teacher will say, Oh, that's just experimental error, and then give you an equation to show you what the answer is supposed to be.

Generations of students (including you) were educated to believe that nature provided neat, clean, graph-able answers. That's mostly because teachers need neat, clean graphs in order to be able to grade tests. The truth about nature is far messier and more interesting.

Looking back at Ed Lorentz' example, you can see that although two very similar storm-system computer simulations start off giving us pretty much the same weather, before too long the lines separate (diverge) and begin to follow different paths. After a short time, the weather systems give entirely different weather altogether.

My friend Huw Davies, physics professor at the Institute for Atmosphere und Climate at the Swiss Federal Institute of Technology in Zurich (Huw specializes in predicting weather patterns using computer models), told me in an email, "Forecasts are now often of high quality out to 3 to 4 days. However, the large-scale atmospheric flow is inherently chaotic in nature and the 'curse' of growth of the initial error hits the forecasts at around this time period (and sometimes much sooner!). Thus weather services now try to estimate the reliability of the forecast—in effect, forecast the accuracy of the forecast!"

What does that mean? Tiny variations in initial weather conditions can change the weather radically, which means that weather is essentially unpredictable outside of three or four days. The best job to have in life? Weather forecaster. You spend all of your time taking educated guesses, you never have to be right, and they pay you anyway. What a deal!

A Snowstorm From Hell

We had a snowstorm in Colorado back in October 1997 that illustrates it beautifully. One Friday morning, we all got up to go to work to hear on the radio that a storm was expected to drop, oh, 3 or 4 inches of snow. No big deal. At maybe 10 o'clock, they were saying, oh, 5 or 6 inches. At noon, it was, oh, 7 or 8 inches. We weren't nervous. We live in Colorado. Snow is just snow.

At about three that afternoon, I thought I might as well head home. It was Friday and it was a good excuse to leave early. I felt a little guilty and wimpy. I mean, getting run out of the office by a little ol' piddly nothing of a storm. I made it home just fine.

My neighbor Scotty left his work at five, two hours later. He didn't make it home for a while. The storm, the severity of which all of the forecasters missed completely, slammed into Colorado and shut our city down for five days. Scotty got partway home, slid off the road into a ditch in a blinding snowstorm, and barely was able to hike over to some stranger's house where he stayed for the next three days. Snow dumped from 2 to 4 feet all over us, with drifts from 4 to 8 feet. Roads vanished, people froze to death in their stuck cars, hundreds of miles of freeway were closed, the airport was shut down for days, the Denver Broncos had to hitchhike to the airport to get their charter flight to play an away game. It was a mess. A huge, unpredictable, unpredicted mess.

Why did they miss it? Easy answer: tiny little variations in the weather made it impossible to predict. As CNN wrote about it, "Ironically, the storm, which hit the Utah mountains Friday, did not appear to have extraordinary intensity as it crossed over the Rockies."[15]

A Hurricane That Fizzled

A similar storm happened in the Gulf in 2002, but in the opposite way. Here's the excerpt:

MONTEGUT, La. (AP)—Hurricane Lili gave Louisiana's coast a 100 mph battering Thursday that swamped streets, knocked out power and snapped trees. But residents were thankful it was not the monster they were warned was coming.

84

More than a million people in Texas and Louisiana had been warned to clear out as the hurricane closed in with terrifying intensity. But in an overnight transformation even forecasters could not fully explain, Lili weakened from a 145-mph, Category 4 hurricane to a Category 2.

And after its center crossed land at Marsh Island, the storm's winds dropped again, falling by midday to 75 mph, barely a hurricane. Instead of a potentially catastrophic 25-foot storm surge, more manageable surges of 6 to 10 feet blew in.

"A lot of Ph.D.s will be written about this," said National Hurricane Center Director Max Mayfield.

Mayfield, who predicted Lili would end the day as a 39-mph tropical storm, was at a loss to explain the hurricane's fluctuations in the Gulf Mexico. While the colder waters in the northern Gulf might explain a weakening of the hurricane, they do not account for why it had gained strength so dramatically earlier in the night.[16]

Bingo. The experts thought a big monster of a hurricane was coming, but for reasons they couldn't even figure out afterward, it just sorta faded away. Some butterfly flapped his wings a little less energetically than he might have. Maybe.

A Treacherous Slope

Another excellent and weather-related example of chaotic collapse waiting to happen is personally important to me. I learned to ski in high school in Switzerland and have continued to ski since in various states and countries. I like to do a little back-country every now and then (you know, away from the slopes and—how to say this politely?—poseurs), and when you get off the slopes and into the trees or chutes, then avalanches begin to have more than a theoretical existence.

I tried for years to figure out how to explain it well, but this article I got out of *Wired* magazine says it perfectly:

Any snowy slope steeper than 35 degrees has the stability of a fleet of eighteen-wheelers suspended by fishing line over a layer of bowling balls. An ounce too much pressure in the wrong spot at the wrong moment and the entire structure comes down. . . . Snow is just so complex in terms of the processes affecting its structure. When I worked on the Apollo space program, I thought rocket science was the hardest form of physics, but snow science is even harder. We always knew tiny differences in weather make huge differences in snow-pack stability, but we could never pinpoint what happened. Now we can actually see the bonds changing between individual grains.[17]

You can see the chaos language emphasized in the article—"an ounce too much pressure" (like a butterfly landing on the snow, maybe?) and "tiny differences in weather make huge differences"; that's chaos in action. The flapping of a butterfly's wing can destroy a village and kill people, or at the very least make my ski day a lot less than pleasant.

Two Trees, Two Blackouts

Speaking of avalanches, here's a different kind of avalanche that affected the lives of millions of people all in a single instant, from *New York Times* On-Line:

When an overheating electrical transmission line sagged into a tree just outside Cleveland at 3:32 p.m. on August 14 [2003], the events that would lead to the greatest power failure in North American history began their furious avalanche, according to the most extensive analysis of the blackout yet.

The failure of that transmission line was crucial, because it put enormous strain on other lines in Ohio. Soon a utility that serves southern Ohio, with its overloaded lines close to burning up, sealed itself off, creating in very real terms an electrical barrier between the southern part of the state and the northern.

What happened next, by this account, was almost inevitable: To the north, Cleveland, starving for electricity, began to drain huge, unsustainable amounts of power from Michigan and then Ontario, knocking out more lines and power plants and pushing the crisis to the borders of northwestern New York.

First the New York system, acting to protect itself, sealed the state's border with Canada, the analysis found. But that only created a different, devastating problem: New York power plants, without anywhere to quickly send electricity not needed within the state, overloaded their own system. That in turn quickly led to a general shutdown—the last stage in the largest blackout in the nation's history.[18]

Notice what happened: a power line hit a tree branch in Ohio. One power line. One tree branch. But it set off a cascade of events that had a devastating and curious result—because New York had too much power and nowhere to send it, suddenly everyone in a large part of the state had no power at all!

One power line. One tree.

The same thing happened two weeks later in Italy:

A storm-tossed tree branch that hit Swiss power lines helped trigger a massive blackout in almost all of Italy on Sunday . . . Initial investigations indicated a chain reaction that started in Switzerland and moved through France. In Switzerland, a tree branch hit and disabled a power transmission line. This caused another Swiss line to overload, which then knocked out French transmission to Italy.[19]

One tree branch. One power line. In Switzerland. Through France. And it wiped out all the power in Italy. Just like the storm that shut off power to Jurassic Park, just before people stopped *having* lunch and started *being* lunch. It's a good thing that neither New York nor Italy has free-range dinosaurs.

"We're All Clueless"

You can see how the weather can affect all of life in dramatic and unexpected ways: we almost expect the weather to do the unexpected every now and then, since we've seen it happen so many times before. It might disturb you to know that the economy is probably more unpredictable than the weather. In fact, after weatherman, the best job

87

to have might be economist. Here's a quote from the *Rocky Mountain News* in Denver: "James Paulsen, the chief investment officer for Wells Capital Management, admitted a dirty little secret that the economic pundits rarely have the guts to share: They can't always predict the future. *'We're all clueless,'* he exclaimed, confessing that in this new economic era, old rules for forecasting didn't seem to be working."[20]

Now let me just say that this article isn't written all that well. There's a world of difference between "they can't always predict the future"—I mean, that's not news; we don't expect anyone to be able to predict the future all that well—and "We're all clueless." "We're all clueless" is not the people you want to be giving your money to for them to invest for you so that you can retire and go live on an island somewhere without dinosaurs for the rest of your life.

Dave Berry says it well:

Whatever you do, do NOT put money in the stock market. The reason you should avoid the stock market is that—to put it in technical terms—nobody knows anything. This is abundantly obvious from the financial reporting on the TV news. No matter what the stock market does, the TV news always boils down to this:

Tom Brokaw: The stock market today went either down or up, and nobody on this Earth knows why. For more, here's our financial expert.

Financial Expert: Tom, analysts attributed the movement of the market to a market movement, in which the market moves either upward or downward, although sometimes it holds still.

Brokaw: And this is expected to continue?

Financial Expert: Tom, it's too soon to tell.'[21]

"We're all clueless." Doesn't that just make you want to run out and give them all your money? What a great job to have! Again, all you really have to do is to tell people why you were wrong.

Local Maxima

It's even more frustrating when Chaos Theory tells you that the best horse doesn't always win the race. Since tiny variations in initial conditions can completely change the outcome, sometimes those tiny variations give results that aren't "optimal," that is, not really what is best in the long run. I kinda think that's how we came up with the knee, which, let's be honest, may be great for bending and walking and stuff, but when it comes to skiing back-country or the bumps, or playing soccer or volleyball, we really could just have used a better design.

The rules of Chaos tell us that tiny variations cause things to track toward what we call "local maxima," not necessarily towards the best result. The best way to understand that is to think of going mountain climbing. You look around to pick the tallest mountain—There it is, over there! Let's climb that one. And when you get to the top and look around, you see taller mountains all around you that you couldn't see from your starting point. Your initial conditions aimed you up a pretty good mountain, but not the tallest. That's what seemed like the *maximum* height, but it turned out to be the highest mountain only in a local sort of way, that is, it's the highest mountain that you found when you looked around.

Chaos aims you up a mountain, but sometimes when you get there, you don't have the best knee you might have had, that is, not the highest mountain, just a high mountain. You just have a pretty good knee. (I've had four knee surgeries, so I'm into this bad knee thing.) Chaos didn't ask the knee designer about skiing and soccer and volleyball, just about running after wooly mammoths and away from saber-tooth tigers, which knees are probably just fine for doing, since if you lose the race, bad knees are the least of your problems.

Beta and VHS

There are some recent examples. One is the battle between Betamax and VHS for the video world. Beta had a better product—VHS won the war. Why? Chaos Theory explains it in Chaos Theory terms; tiny variations in initial conditions made the difference. That is, slightly more people bought VHS systems as people were investing in home systems, and as time went on, it became easier and easier for the market to swing toward VHS rather than Beta. So Beta is long gone, and VHS is here, at least until DVDs wipe it out—the next stage in the process, and it's happening pretty fast.

Apple and Mac

Something similar happened between Apple and Microsoft.

Let's be honest: Apple has a better operating system. Apple has always had a better operating system. It's so good that Microsoft felt the need to copy the whole thing and call it Windows, but you may remember that when Windows 95 came out, they called it Apple 89; that is, Microsoft was six years behind Apple in designing the same thing. Apples crash less often, they are more user-friendly, they're cooler looking, they're clever, and they are just better in every way.

So why am I typing this on a PC? Why are most of us working in a Windows environment? Why is Bill Gates the richest person in human history? Why do more than 90% of the world's computer owners use Microsoft software?

Without going into detail, way back at the beginning Microsoft and Apple, both made decisions in the way they were going to market and make their software available, and it worked for Microsoft and not for Apple, at least not nearly as well. It was not a case of dividing up the market; it was sheer market dominance, and Apple survived by the skin of its teeth because of brilliant design innovations. Simple decisions at the beginning that clearly had unpredictable futures—otherwise Apple would have made the same decisions—had a huge impact not just on Bill Gates and Microsoft, but on the entire planet.

For example, if there were multiple operating systems out in the world of computers, then computer viruses, just for starters, would not be nearly as threatening to the very stability of the world's economies. Most viruses are designed to be aimed at Microsoft systems. Apples are seldom targeted. The very real threat exists that the world can be brought to its poorly designed knees by an effective and clever virus that can shut off or damage our electricity grid, the phone system, the Internet and the World Wide Web, and who knows what else? All because of tiny little variations in a series of decisions made by Bill Gates years ago.

Cloning and DNA

Chaos Theory applies to far more fields of human inquiry than weather and economics. One is the field of cloning. Here's a picture of Rainbow, the donor cat, and CC, the clone:

www.mun.ca/biology/ scarr/Cloned Cat.htm

It doesn't take a genetic scientist to see that CC and Rainbow don't look anything alike. Clones are supposed to be, well, clones, you know, things that are exactly alike in every way. I mean, at the very least they are supposed to look alike, aren't they?

I guess not. These two cats don't look alike, and when CC grew up, it still didn't look like Rainbow.

Why doesn't cloning give clones, I mean, you know, things that look alike? To quote from the *New York Times* special DNA edition:

DNA may be elegant, but it often has been accorded far greater powers than it possesses. With all the breathless talk of human DNA as a grand epic written in three billion runes, the scientists complain that an essential point is forgotten: DNA, on its own, does nothing. It can't make eyes blue, livers bilious or brains bulging. It holds bare-bones information—suggestions, really—for the construction of the proteins of which all life forms are built, but that's it. DNA can't read those instructions, it can't divide, it can't keep itself clean or sit up properly—proteins that surround it do all those tasks. Stripped of context within the body's cells, DNA is helpless, speechless—DOA. By the same token, cells need their looping lanyards of genes and would grow as dull as hairballs without them.[22]

All of that is to say that DNA works in a complex, chaotic environment; not chaotic in the sense that there is no order, but in the sense that it is unpredictable—there are too many tiny variations to be able to control or predict the outcome. It's not just DNA; it's the cell, the proteins, the amazing complexity of the process of cellular reproduction, all of these and many more influence the process in unpredictable ways.

That is to say that it's not completely unpredictable. And that's chaos. We know that, if everything goes reasonably well, when we clone a horse, we still get a horse. When a weather system is changed in some tiny way at the beginning of the events that unfold, we still get weather. When the economy or the stock market change unpredictably, we still get economics and stock prices. The results are always inside the range of the possible, but where we land in that range is hard to control or predict. We know we're going to get weather, but we don't know whether we'll get a blizzard, a hurricane, or clear and warm.

There are some more sobering examples of chaos in action that we need to consider.

Questions:

1. Define "Butterfly Effect".

2. Does Chaos imply that there is no order?

3. Is talking about Chaos Theory a really effective way to impressive a pretty paleobotanist if you have gold chains in your chest hair?

4. Why are things unpredictable?

5. About how many days in advance can the weather be predicted under normal conditions? Why can't it be predicted further out in time?

7

The Downside of Chaos

On January 28, 1986, the Space Shuttle Challenger exploded just over a minute into its flight. On February 1, 2003, the Space Shuttle Columbia exploded just a few minutes before it was scheduled to land in Florida after a successful flight. All the astronauts were killed on both flights, and the nation was traumatized each time. As with most of our national disasters, the search to find a cause (and someone to blame) started intensely and immediately.

They really didn't need to bother, although the details were interesting and terrifying. The answer is found in Chaos; there are always too many tiny little variables to be able to control in a high-energy event like a space launch or landing.

The Challenger

On the Challenger inquiry:

From the outset the commission confronted evidence showing that NASA officials ordered the launch to go ahead despite safety warnings. There were charges that the White House had intervened to prevent further delays in the launch so that it would coincide with Reagan's State of the Union speech to Congress set for that evening. NASA had submitted to Reagan a paragraph to be included in the speech saluting McAuliffe.

The blasting of human beings into Earth's orbit aboard an explosive-laden rocket at enormous speeds is an incredibly complex and inherently risky undertaking. The destruction of the Challenger involved one of the largest non-nuclear explosions in history, the equivalent of nearly 1000 tons of TNT.

The apparent nonchalance of NASA officials about safety was appalling. In the final analysis it reflected pressures to maintain an impossibly ambitious launch schedule set by the military, which saw the shuttle as the cornerstone of Reagan's "Star Wars" program, as well as pressures from NASA's corporate clients.

Budgetary and political pressures affected the ultimate design of the shuttle. Following the successful moon landings the NASA budget had been under steady attack, resulting in pressures to lower design standards in order to cut costs. In order to justify its budget, NASA had to demonstrate the space program's military value. This required design modifications affecting the safety and landing capabilities of the space vehicle.

What we see so far is Chaos Theory in action: many variables working together to get the launch going. Some of the pressures were political, some bureaucratic, some budgetary. But the thing that ultimately caused the explosion itself is, again, brilliantly and terrifyingly chaotic:

The Challenger commission soon established the immediate cause of the disaster, the failure of the rubber-like O-rings joining the sections of the solid rocket boosters. [physicist Richard] Feynman established that the O-ring failed due to record cold temperatures at the time of the launch.

To dramatize this, the physicist dipped an O-ring into ice water during a televised session of the commission. The O-ring immediately became brittle. The lack of resiliency of the O-rings at relatively low temperatures prevented them from sealing properly, thus permitting hot gases to escape, resulting in the emission of a flame from the side of the booster. The flame caused the main external fuel tank to explode 73 seconds into the launch.[23]

It was the Butterfly Effect: a tiny little rubber "O-ring," something that had functioned well at normal temperatures in past launches, failed. The temperature on the launch pad on January 28, 1986 was cold, in the 40s. There was concern that it was too cold, but the political pressures were higher for this launch. There was a schoolteacher on board, the first teacher and the first civilian on such a launch. The entire country was watching in fascination; one of us normal people was going into space. Schoolchildren around the country were glued to their TVs, which in a Chaotic way made the disaster that much more profound.

A little bit of rubber, a little bit of cold temperature, a little bit of pressure to launch rather than wait, tiny little variations that had a huge impact on the lives of the astronauts and their families, and because of one school teacher, the event had a huge impact on the lives of children around the country and the world.

The Columbia

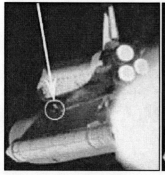 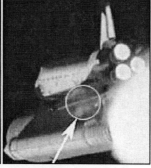

The Columbia inquiry: from the *New York Times*:

The event that doomed the shuttle may have taken just two-tenths of a second, with the impact 81.86 seconds after liftoff.

That is how long it took a chunk of foam to cover about 60 feet, between the spot on the external tank called the bipod ramp, up near the Columbia's nose, and the panel on the left wing where the chunk hit and probably punched a hole.

The shuttle was at 65,860 feet, and was already traveling 2.46 times as fast as the speed of sound.

That was on Jan. 16, about 10:40 in the morning, almost 16 days before the unsuccessful re-entry.

Another way to look at the disaster is that it began almost 27 months earlier: 805 days, to be exact, when the external tank was finished, at a factory in Michoud, La., near New Orleans.

Subsequent analysis found that the processes for testing and inspecting the bipod ramp foam "were not designed to be capable of rejecting ramps" with fabrication problems that would threaten the orbiter.

Or maybe the accident began 20 years before that, when the Columbia first flew, or even a few years earlier, when the shuttles were designed with a newly conceived location for the equipment that would bring the crew home: beside the propulsion system instead of above it, and thus vulnerable to debris.

Maybe it was something in the first 81 seconds of flight. At 62 seconds and 32,000 feet, the orbiter experienced wind shear, as it passed through an area where the wind was blowing in a different direction.

As usual, the external tank fuel had shrunk before liftoff because the fuel was so cold, and that put strain on the bipod; possibly the wind shear helped knock the foam loose, though such shear is common in shuttle launchings.[24]

A piece of foam that a small child could lift with one hand, hit you in the head with and cause no damage. But the speeds were high, and it hit just at the right angle in the right spot, or rather, at the wrong angle in the wrong spot. They'll never know how it hit the wing exactly—the foam was moving at about 500 miles an hour and rotating at 18 times per second. They'll never know whether the foam put hairline cracks in the wing or punched a 10- to 16-inch hole in the wing—in the tests engineers conducted afterward, both occurred. Also playing a possible role—the design of the shuttle, 20 years earlier, 20 years without a problem like this, a problem just waiting for right set of

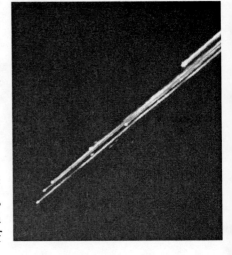

tiny little variables to combine in what the *New York Times* described on February 3, 2003, as a "cascading series of failures that destroyed the shuttle Columbia." A *cascading* series of failures. The Butterfly Effect—a series of tiny events that result in Chaotic collapse.

The next and final example of Chaos alive and well in our world today is both brilliant and terrifying, a brilliant and on-going (at the time of this writing) example, a terrifying study in how tiny actions can affect us all. It happened on September 11, 2001.

September 11

This may be quite a transition for some readers, seeming to shift from science to history or politics. So we need a bit of a disclaimer by way of introduction.

First, Chaos Theory is not limited to physics or even to science. Just as Einstein's earthshaking discovery of a simple equation, $E=MC^2$, turned into earth-shattering events over Hiroshima and Nagasaki and subsequently altered the history books and the direction of world history with the arms race, the Cold War, Mutually Assured Destruction, Détente, and the changing of the very fabric of human existence, so too does Chaos Theory, with its beginnings perhaps in science but also in history or culture, have the potential to impact the way we live our lives. It is a natural, inevitable process to go from tiny variations to earth-shattering, history-altering events. It is, as Chaotic Ian said, the essence of Chaos to do so. What happened on 9/11, with simple roots somewhere back in time, has been earth shattering and history altering. Chaos Theory is no stranger to history.

Second, the process of the changes effected by Chaotic events goes in either of two directions—it dissipates quickly and has no long-lasting effects, or it dissipates much more slowly and the effects may be permanent. There is no real need for the effects to dissipate at all, though most times they do. Einstein's musings on a stool in the patent office in Bern, Switzerland, changed the world in a radically unpredictable way that certainly seems permanent—the changes *may* not be permanent, of course, but it would not be unusual for Chaos to produce permanent changes. The events of 9/11 will have long-reaching, perhaps permanent impact on world history—whether they are permanent will of course be hard to say.

Third, because the impact of 9/11 remains so much with us in society, culture, politics, economics, and in virtually every realm of life (which is of course what Chaos does—it affects every arena of existence), some of what will be written about here may seem to be political, trying to make one side or the other look bad in some way. Please understand that we are writing about the impact of 9/11 on a broad front of human existence because that impact is real. What follows here are the results, radically unpredictable but incredibly real and powerful, of the events of 9/11. We are not interested in talking about mistakes that may have been made, clues missed, or decisions unmade, nor in placing or assigning blame to anyone, not even to the terrorist pilots themselves. It's not that we are carefully remaining morally and ethically neutral—it is that we are simply talking about the results of an event that caused a cascading avalanche of other events in a terrific and terrible example of pure chaos in action.

Some will say that the acts of terrorism on 9/11 were somewhat larger than the flapping of a butterfly's wings, that is, they were hardly small and potentially inconsequential. That is both partially true and historically shortsighted on several fronts. First, the events of 9/11 did not start on 9/11—they started years, decades, perhaps even centuries earlier, with many subtle influences involved in creating the tensions in the Middle East that tend to dominate world politics these days.

These horrific events comprise a continuing cascading series of events that find their roots in tensions in the Middle East, in the Holocaust and WWII, in the years between the wars and the treaties that were signed, in the culture of various European cultures, in the history of the Jewish people in and outside of Europe, in the curious concentration of oil in the Middle East and in the western world's dependence on that oil, in related decisions by the automobile industry, particularly in the United States, to dismantle public transportation in North America and sell lots of cars, even perhaps in our independent nature that moves us in subtle ways toward a desire for automotive travel rather than train travel (which might itself be related to the Cold War and the tension between Capitalism and Socialism/Communism), in America's apocalyptic view of Israel and the Middle East, and on backwards in time in a variety of directions. It was a complicated and unpredictable mix that coalesced into the events of 9/11/2001.

It is also true that we blow up buildings all the time—Hollywood does it to make movies, Las Vegas does it to get rid of old hotels (so they can build new hotels), sports teams do it to demolish old stadia to build new ones, and so on.

Further, there have been any number of buildings that were bombed that didn't change world around us—I drove by the US Embassy in Nairobi a year or so after terrorists blew it up, with negligible impact on life in North America, similar to that felt by the bombing of the embassy in Tanzania the same day. Buildings are blown up all the time in Israel and Palestine, and in the many other wars that rage around the globe constantly.

There was something unusual about 9/11. I submit that the bombing of the Pentagon was not significant in the way that the destruction of the World Trade Center towers was. A large part of that significance may have been found in the fact that we all watched it happen live on television. We only saw the Pentagon smoking later, not a plane flying unbelievably into its side. We watched the first tower burning, we saw the plane fly into the second tower, we watched the people fleeing, we saw the bodies floating gently and horribly in the air on their last journey of blind panic to the ground below, and we watched the towers settle into themselves, dissolving almost magically into nothingness as they collapsed into the graves prepared for them by those who dug the original foundation.

We saw endless repeats of planes flying into buildings blowing flaming bits of human detritus into vast airscapes of horror. We wondered in pain what it meant for the world. It is a pain that seems to have aftershocks larger than the original. We are not alone. You all know the pain. It kept you awake at night with Steven King stories haunting your prayers, but not fictional, not movie images imprinted on our dullish brains, but real thundering collapsing towers of death and destruction, and exploding holes that swallowed planes and people as though they never were. It is and remains a nightmare.

Was that the catalyst for the cascading series of events that followed, that is, just the fact that we watched it all happen? We'll never know if the world would have changed so much if we hadn't seen it live on TV, but it seems likely that TV has become one more flapping of the butterfly's wing, one more effect that drags us further into Chaos.

In the years since, I have watched for that Chaos, collecting stories, headlines, articles, and photos. It should be no mystery to any of us that the effects of 9/11 continue to cascade, reinforced by other factors—all one has to do is to look at the headlines to find the word "Iraq" to know that it continues. Regardless of one's views or opinions about Iraq and whether we were or are justified in being in that country because of 9/11, it is clearly and historically true—we went to war with Iraq because of 9/11.

The immediate effect of the collapse of the two towers was easy to see, though in some cases only if one was watching for it. Airports and other forms of transportation shut down country-wide, and then worldwide. We lived in fear for a day and beyond, even until now.

Let's look at some headlines and articles since. Many of these are from Colorado area newspapers, since that is my home:

ATTACKS SHOOK THE WORLD ECOMONY (Denver Post)

NIGHTMARE ON WALLSTREET ((Denver Post, SEPT 22, 2001)

RECESSION NOW APPEARS CERTAIN (Denver Post)

1.8 MILLION JOBS VANISH POST-SEPT. 11 (Denver Post)

U.S. ECONOMY IN WORST HIRING SLUMP IN 20 YEARS (FEB 03) **(Denver Post)**

JOBLESS RATE HITS 6.4%, HIGHEST LEVEL IN 9 YEARS (JULY 4, 2003) **(Denver Post)**

AIRLINES MAY HAVE LOST $15 BILLION (FEB 6, 2002) **(Denver Post)**

WAR COULD SEND MAJOR AIRLINES INTO BANKRUPTCY (Denver Post)

The world airline industry lost more in the past two years than it made in profits in the combined previous 45 years. (Jan 11, 2003)

93,000 JOBS LOST IN AUGUST (2003) **(Denver Post)**

Concerns raised some jobs may be gone forever

SEPT. 11 ECONOMIC TOLL: $1 TRILLION (SEPT. 7, 2003) **(Denver Post)**

The attacks of 9/11 and the fallout of the last two years have cost the world about $1 trillion - that's a thousand billions - and half of that has been borne by the United States. Sept. 11, 2001, effectively set off a non-nuclear World War III, said David Littmann, chief economist for Comerica Bank-Detroit.

REFUGEES FACE TOUGH TIME GETTING INTO U.S. (Denver Post)

A NEW ARMS GAME—(Newsweek, Dec 24 2001)

In one stroke last week, President George W. Bush called an end to a 30-year era of arms control treaties . . . yet Bush came away unscathed politically from what once was a hot button issue—with the war on terror, "who's going to complain?"

FED SILENCE SAYS LOTS (MAR 22, 2003) (Denver Post)

(Alan) Greenspan seems to have 'no idea where the economy is heading'.

'Because of the unusually large uncertainties clouding the geopolitical situation in the short run and their apparent effects on economic decision-making . . . the committee decided to refrain from making a determination until some of those uncertainties abate.'

CHILD CARE SPENDING A WAR VICTIM (Denver Post)

TERRORISM OVERSHADOWS ENVIRONMENT, GOODALL SAYS (Denver Post)

AT-RISK YOUTH EFFORT UNLIKELY TO BE REVIVED (Denver Post)

BILLIONS MORE FOR DEFENSE (JAN 24, 2002) (Denver Post)

LOSSES, BEFORE BULLETS FLY, BY NICHOLAS D. KRISTOF, NY TIMES ON-LINE, MARCH 7, 2003

So let's take stock of how our invasion of Iraq is going. The Western alliance is ferociously strained, NATO is paralyzed, America is resented by millions, the United Nations is in crisis, U.S. pals like Tony Blair are being skewered at home, North Korea has exploited our distraction to crank up plutonium production, oil prices have surged, and the world financial markets have sagged. And the war hasn't even begun yet.

HOW MUCH WILL THE COUNTRY SEPEND ON WAR ON TERRORISM? (©*Daily Local News 2003*)

Simply put, the United States is spending more than $99,000 a minute while occupying Iraq. That's nearly $4 billion per month. And President Bush declared *his intention to ask for $87 billion more during his address to the nation Sunday night. It's another $1.1 billion per month in Afghanistan.*

FOREIGN VIEWS OF U.S. DARKER SINCE SEPT. 11
(September 11, 2003), World Opinion, NY Times On-Line

In the two years since Sept. 11, 2001, the view of the United States as a victim of terrorism that deserved the world's sympathy and support has given way to a widespread vision of America as an imperial power that has defied world opinion through unjustified and unilateral use of military force.

TERRORISM BLAMED FOR DEFICIT IN BUDGET— QUICK CHANGE: SURPLUS NOW $*106 Billion* DEFICIT
(January 24. 2002 Washington Post) (ed.: the deficit finished 2002 at *$159 Billion*)

US DEFICIT TO TOP $*200 Billion* IN '03 (NY Times) (forecast *$300 Billion* in '04)

THE PRESIDENT'S BUDGET PROPOSAL (February 4, 2003, Tuesday, NY Times Online)

Pres Bush sends Congress $2.23 trillion budget with record deficits, budget forecasts deficit of $304 billion *in current fiscal year (2003) and* $307 billion *for fiscal 2004; projects* **$1 trillion-plus deficit over next five years;** *includes no projection of cost of any war with Iraq.*

U.S. DEFICIT SEEN AS RISING FAST (NY Times On-Line March 5, 2003)

The government's 2003 shortfall could soar to $400 billion if Mr. Bush's tax cuts are approved and if war costs this year run into the tens of billions of dollars.

WHITE HOUSE SEES A $455 BILLION GAP IN '03 BUDGET (NY Times, July 16, 2003)

The White House today projected a $455 billion *budget deficit in the current fiscal year (2003), by far the government's largest deficit ever and $150 billion higher than what the administration predicted just five months ago.*

WHITE HOUSE TO PROJECT DEFICIT OF $521 billion IN 2004 (WASHINGTON [Reuters])

The White House will project that the federal budget deficit will peak this year at $521 billion, shattering previous records. (As it happened, the deficit was much less than projected -- $375 B for 2003, $445 for 2004 -- af)

U.S. DEBT GROWTH IN ONE YEAR DRAMATIC—(Denver Post) 10/8/03

The U.S. government has run up as much debt in a single year as it did during the country's first 200. *The total national debt grew by $576 billion during the 12 months that ended Oct. 3, according to the Treasury Department's Bureau of Public Debt. The total national debt, which includes what the government owes the public and itself, rose from $6.23 trillion to $6.81 trillion during the past 12 months. That works out to more than $23,000 per American, or enough to buy everyone a 2004 Chevy Blazer or Ford Taurus.*

Let me say again that there is no hidden political agenda nested in the headlines I've chosen—these are simply the direct and indirect results of the events of September 11, 2001, along with some critical elements that preceded 9/11 and many decisions made afterward that took the country and the world in certain directions. The Butterfly Effect is not about the mere flapping of one butterfly's wings—it is about the cascade of events that is set off by that flap, all of which reinforce the energy of the avalanche that results.

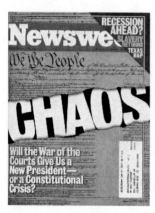

There are a couple of interesting events that were critical to post-9/11. First was the Chaotic election of George Bush. If you recall, it started with a few ballots in Florida that didn't get registered clearly—the "hanging chads," little bits of paper holes, like a donut hole, that didn't quite get pushed through by the voters. Normally, little bits of paper would have no impact, but it turned out that the vote was close in Florida and those votes mattered. It also happened that Florida's electoral votes were pivotal, and when the dust cleared, Bush won the state by a few votes, and won the national election by . . . one electoral vote. But before that, he was given the election in Florida by the US Supreme Court, and that vote itself was decided by . . . .one vote. Some truly tiny bits of paper, a few votes, one vote here, one vote there, and the course of American history was changed. Though each party has its view of what might have happened after 9/11 had Al Gore been president, it is clear that it could have been radically different than what did happen, for better or worse.

Add to that some marketing decisions made again by Bill Gates and Microsoft to bundle their browser software with their operating system, the subsequent action by the government directed toward Microsoft, which in turn caused the (overheated and highly over-inflated) high tech market to crash in March of 2001, dragging the rest of the market and the economy along with it, and the stage was set for the events of 9/11 to have a much larger impact. Had the high tech market not collapsed, then much would have been different with the economy, though the wars in Afghanistan and Iraq would very likely (though not absolutely) been pursued. Had different decisions been made with respect to both those countries, then the economy might have been affected differently.

But clearly a number of things would have been much the same. The airline industry was battered and still teeters on the edge of collapse—it lost as much in two years as it made in the previous 45 years combined. We still, even in 2005, could lose United, American, Delta, US Air, Northwest, and Continental airlines, and that's not to mention air carriers from other countries that might be affected similarly.

Immigration into the United States was on the verge of being dramatically restructured, something that only in 2005 is being discussed again, and not nearly in the same terms as on September 10, 2001. This will hurt poor families from all of those countries from which our immigrants normally come, and may in turn drive up prices for fruits and vegetables in US markets if labor gets tight and wages rise.

Refugees will continue to struggle to gain access to the US from around the world, and that has a powerful impact globally, not least because it can put pressure from the poor and dispossessed on governments and economies, exacerbating poverty, the possibilities of insurrection, AIDS, SARS, famine, civil wars, and so on.

With the border crossings into Canada and Mexico so tight, industry in the United States has been significantly challenged, especially with regard to what is called "just in time" inventory maintenance. This is where industries on the borders, particularly the border between Detroit (the automotive industry) and Windsor, Ontario, can no longer have parts delivered to be put directly onto the assembly line rather than being stored in expensive warehouses—when border crossings take hours or days longer, assembly lines can't wait, and so costs go up.

Add to this the overbuilding that industry experienced before the stock market tanked in anticipation of a greatly expanding economy— suddenly sales dropped away, and it was necessary to find a way to sell an overproduction of goods that had to continue to be produced— why do you think that the car industry sold us cars for nothing down and no percent? It wasn't because they love us—they simply had to sell the cars that they had to continue to produce.

There may be social consequences—when "child care spending" goes down, then families can be impacted in negative ways. Parents are forced to quit jobs to stay home with kids, or kids more often are at home alone, increasing the chances for drug, alcohol, and sexual experimentation, as well as potentially hurting academic success rates. Put that together with reduced funding for "at-risk kids," and the chances for more crime increase, with more gang activity (as gangs continue to replace the family in some communities), and we need more jails, more trials, more expense.

With the focus of the nation shifted away from environmental issues to the war on terror, we may see an erosion of our commitment to improving the environment, so we may have more pollution, more smog, and more pollution-related diseases. With states hurting for cash, our local infrastructure may erode—roads, police and fire departments, schools, poverty programs, intervention programs for battered women and abused children, and so on.

With respect to employment, the neat and awkward dovetailing of 9/11, the collapse of the high tech market, the out-of-control health care industry, and the acceleration of globalization brought on by the cheapness of international telecommunications (fiber optic cables) has meant that thousands if not hundreds of thousands or more jobs have moved permanently overseas to countries where labor (salaries, medical benefits, pensions, etc.) is far cheaper. It is a process that probably would have happened anyway, but much more slowly and perhaps in such a way as to allow the economy and its workers to adapt better rather than having to deal with the catastrophic collapse of the jobs market.

Add to this some subtle influences in modern Asian history (the Chinese economic dragon has finally awakened, India was sitting on an educated labor force ready to take over the tech-service industry, and other countries such as Malaysia and Thailand were just coming out of the Pacific economic collapse), and the stage was set for truly profound disruption on a massive scale, just waiting to be catalyzed by something like 9/11.

Now we have the US, with a much smaller tax base and far larger military expenditures, spending its way into deep deficits with the largest national debt in history, focusing its energies on non-income-producing industries like the military and the war on terrorism, siphoning R&D money away from more productive industries, while other countries are taking advantage of cheaper labor and lack of workplace oversight to build their economies and provide jobs and industries for their people. There may be a truly massive and historic shift away from a US-centric planet—our days of primacy may be on the wane as the orbit of economic power shifts to Europe and most significantly to Asia.

But a good chaotician knows that Chaos renders the universe unpredictable, and there are too many butterflies still flapping out there to be able to say where it will all end up.

Plus, there is a flip-side to Chaos that can take all of this bad news and turn it into good. But not in any way that you might predict or anticipate.

Questions:

1. What was the direct cause of the Challenger disaster? What were some of the indirect causes?

2. How was the Columbia disaster similar?

3. On September 11, 2001, something less than twenty people changed the course of human history. Is that an example of Chaos Theory? Why or why not?

8

Order: The Upside of Chaos

ChaosChaosChaosChaos
ChaosChaosChaosChaos
OrderOrderOrderOrder
OrderOrderOrderOrder
OrderOrderOrderOrder

The upside is, out of Chaos we seem to get Order.

If you're very good, you will immediately say, OK, bud, what do you mean by the word *order*? It's a good question, particularly in these postmodern times when some are saying that "order" doesn't really exist but is imposed by us on a disordered universe—we just make it up.

Illustrating Order

Kjell B. Sandved, former staff photographer for the Smithsonian's National Museum of Natural History, spent thirty years photographing something like a million butterfly wings, trying to find commonly used letters, numbers and symbols. You can check it out at http://www.butterflyalphabet.com. Ultimately he found all the letters of the English alphabet and the numbers from 1-9.

That's beautiful. But it's not order. It's coincidence. It was an amazing feat of hard work and perseverance, but finding shapes on wings is just a matter of looking for the shapes you wanted to find. You could probably find the Arabic alphabet on butterfly wings, or Japanese hiragana or kanji. Kjell found a wing saying "Hi," one picturing a lady in a shawl, another with a singing group of ghouls, another with long rows of hearts.

They're extraordinary to look at, but it's not order. It's exactly what the post-modern critic is criticizing: the imposition of order on nature, an order that exists only in our minds.

The Elements

However, this is order:

This is, of course, the Periodic Table of the Elements, and it demonstrates that we do not impose order on a disordered universe; the order is there waiting to be discovered. It's mathematical, there are patterns there to be perceived, and the patterns allow us to make predictions.

In the case of the Table, the numbers of particles inside the atom of an element are linear and discrete—we don't find half particles or partial atoms. They start at the simplest of levels (hydrogen) with the simplest of structures (a single proton, a single electron), and it was, as it happens, the first atom to appear in the universe after Big Bang. The elements get more complicated in a structured and predictable way— by adding protons and electrons in a linear fashion—and somehow,

112

amazingly, with nothing more really happening than adding little bits of spinning energy to a nucleus, each addition radically changes the nature of the new element. There's almost something magical about it, and certainly chaotic – the tiniest of changes at the smallest of levels results in radical differences between elements, differences so large that it's hard to understand.

One might also note that in the history of chemistry, chemists have looked at the blank spaces (where there were no known elements) and predicted that they would be there, ultimately filling many of the spaces by deliberately seeking to do so. That is one way we define order: discernible patterns that allow for reasonable predictions that turn out to be true. One might try to say that the patterns don't have any real existence, that we impose patterns on Chaos, but if the patterns are mathematically consistent, then they exist beyond our perception; they have a physical, mathematical reality that translates into real reality, things that you can touch or measure, things that are consistent and, within the parameters of Chaos, predictable. That is, we can predict that the elements will be there and what they will look like at the atomic level. But we have no way to predict what the qualities of the element itself will be.

Beehives

A beehive is a great illustration of order in nature. Each cell in a beehive is a hexagon, a six-sided figure with equal sides and equal angles (again, as equal as Chaos allows). Bees know nothing of geometry, nothing of hexagons and sides and angles, and yet the pattern is unmistakably and wonderfully mathematical. It is also unexpected and surprising: Why not sloppy little caves that have nothing to commend them? How do bees come up with hexagons?

We don't quite have an answer for that question (other than just saying "instinct," which only gives a word to the answer we still don't have), but it's not the only example we have that creates the same question.

Logarithmic Spirals

The graph below comes from the Fibonacci series, which is found by taking the following numbers and doing something not very creative with them. The numbers are 0,1,1,2,3,5,8,13,21 and so on—each number is found by adding the two previous numbers together (0+1=1, 1+1=2, 1+2=3 . . .). You then take each pair of adjacent numbers and make a fraction out of them: 1/1, 2/1, 3/2, 5/3, 8/5, 13/8, 21/13 etc. If you use your calculator to turn these fractions into decimals, they gradually get closer and closer to a certain number called the Golden Mean, also called *phi* (φ), which is 1.6180339887 . . . not all that interesting . . . until you begin to find it in nature. A simple graph of what is called a Fibonacci spiral looks like this:

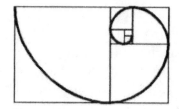

Below is a cross-section of a nautilus shell. The similarity between the spiral above and the shell below is far beyond coincidental. The

nautilus is one of the oldest still-living creatures on Earth, and its shell forms a perfect Fibonacci spiral. The chambers of the shell can be found by using an equation as follows:

$$r = a * e^{\theta \cot(b)}$$

which is almost impossible to get if you're not a math geek, so don't sweat it. To quote from the website where I found it, "The curves of the spiral are called equiangular (or logarithmic) spirals. The smaller the constant a, the tighter the spiral becomes. The constant b is the angle between the radial line and the tangent line. This is consistent with every turn."[25]

114

Here's what I want you to get from all of that pretty complicated explanation—the chambers of a nautilus shell are formed naturally as they have been for hundreds of millions of years. It's a very simple creature, very old. But the life process that produces those chambers uses *logarithmic spirals*. Now that's order happening naturally in nature in an amazing way. And it doesn't stop with obscure should-be-extinct-by-now shellfish.

Sunflowers can be the same, and the same spiral is seen in the way that its little inside flower gizmos grow—petals, I think. Here's what a website says:

You can see that the orange "petals" seem to form spirals curving both to the left and to the right. At the edge of the picture, if you count those spiraling to the right as you go outwards, there are 55 spirals. A little further toward the centre and you can count 34 spirals. How many spirals go the other way at these places? You will see that the pair of numbers (counting spirals in curving left and curving right) are neighbours in the Fibonacci series.[26]

And it's not just in flowers. Water drains in a Fibonacci spiral. Hurricanes spin in Fibonacci spirals. And spiral galaxies spin in Fibonacci spirals.

It's simple to produce. It produces patterns in nature that seem to come from nowhere, or all from the same place, wherever that is, patterns that are immediately discernible and readily measured and reproduced. The pattern occurs in very old things, both living and not, in small things and large, in things that have no physical connection at all, and yet, it is the same pattern, formed from the same simple series of numbers—0,1,1,2,3,5,8,13,21 . . .

What Fractals Show Us

So where do we go with this? First, we have to look at fractals, because Fibonacci spirals are a small, fractalated part of a fractalated universe. (Now, there's a sentence nobody's going to get.)

Here's another one, this one with the definition of *fractal: Term coined by Benoit Mandelbrot in 1975, referring to objects built using recursion, where some aspect of the limiting object is infinite and another is finite, and where at any iteration, some piece of the object is a scaled down version of the previous iteration* (cf Properties of Fractals Discussion, also Plane Figure Fractals Discussion).[27]

We've got to do better than that. A fractal is a graph (OK, bad start, that's a total math-geek word), no, really, it's just a graph that does a couple of things: one, it continues forever while getting infinitely smaller; two, it seems to repeat itself over and over again (we'll call that "self-similarity"); three, you can't predict what it's going to look like at all (see—there's chaotic unpredictability); four, it's really beautiful; and five, it has a really simple starting point. It illustrates one of our fundamental principles of Chaos Theory—tiny little variations in this case resulting in the most complicated things man has ever discovered, none of which can be predicted from the starting point.

OK, we've got to do a little math. If you hate math, you'll hate this. It's math. You could skip this part, I suppose. I mean, how am I supposed to tell? It's not like I'm watching you.

If you have a vague memory of doing graphs back in math class in high school, then you might have a vague memory of what those graphs used to look like. For example, let's take this simple equation:

$$Y = X^2 + 2X + 1$$

Most high school students—OK, some high school students, but no adults other than math geeks—can tell you that if you graph equations of this type you get what is called a parabola, and all parabolas look a little like this:

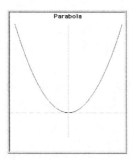

Now we'll take another equation, a little more complicated:

$$Y=X^3+3X^2+3X+1$$

If you graph this one, you get a graph that will look something like this:

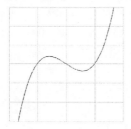

See, isn't this easier than math class? I'm giving you all the answers and there won't be a test.

Here's an even more complicated equation:

$$Y=X^4+4X^3+6X^2+4X+1$$

And when I graph this one, instead of getting some monster hideous graph, I get something like this:

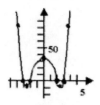

Those of you who are (let's be honest) math geeks are getting all worked up right now because the equations and the graphs don't exactly match, with different maxima and/or minima, y-intercepts, and all of that mess that really you just need to get over. Feel free to go out and graph them all correctly and mail them to me at this address:

1600 Pennsylvania Avenue NW
Washington DC 20050

I promise I'll get right on it.

OK, they're all off graphing things, so the rest of us can get on with this. The point here is that we keep getting much more complicated equations, but the graphs are really just not all that special. Boring, that's what they are. We math teachers have to give you these, though, because we need to be able to have easy answers to grade so that we can give you grades and you can get through school, though everyone graduates pretty unimpressed with graphing as a job option, because we only gave you boring graphs.

Here, check these out—complicated equations, boring graphs.

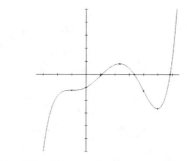

$$Y=X^5+5X^4+10X^3+10X^2+5X+1$$

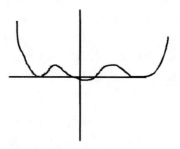

$$Y=X^6+6X^5+15X^4+20X^3+15X^2+6X+1$$

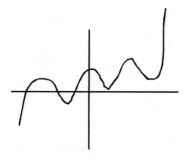

$$Y=X^7+7X^6+21X^5+35X^4+35X^3+21X^2+7X+1$$

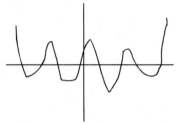

$$Y=X^8+8X^7+28X^6+56X^5+70X^4+56X^3+28X^2+8X+1$$

Simple Results from Complicated Starts

Mostly the graphs just get another lump when they get more complicated. OK, pay attention, here comes another Chaos statement: *We are getting simple results from complicated starting points.* That's not really supposed to happen. Science has always assumed that if you start simple, you get simple results, and if you have complicated results, you must have started someplace complicated.

Chaos Theory tells us that not only do complicated starting points give us simple (and boring) results, but many times simple starting points give us the most amazing results.

So let's take this equation, really a pretty simple equation, and see what happens when we graph it:

$$Y=Z^2 + C$$

We need to define a few terms. (Math geeks, you can come back in the room now. Your graphs are lovely.)

- Z is a complex number

- C is a constant

- Choose initial random values for Z and C

- Y will be a new complex number

- Substitute it for Z and do it over and over again.

- Graph it, point by point, and watch what happens . . .

At first, you just get little blobby bits scattered at random around the place:

28

The points generated by the graph land unpredictably—you don't know where they are going to land or whether they'll group up, float around friendless, or whatever.

But if you let the process continue (using a computer is useful at this point—the early guys graphed these by hand, and then ended up in homes for the mathematically insane), and if the points converge (sometimes they diverge and you get garbage), then the results are fascinating.[29]

And a bit longer…

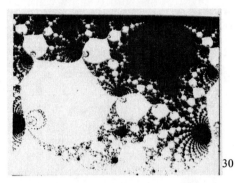

30

Gradually, out of the blobbiness, we see what have been called the most complicated things humans have ever discovered. They are called fractals and they possess some truly intriguing characteristics, including self-similarity, weird dimensionality, and beauty.

Self-Similarity

Here's a little snowman that will help us understand it better:

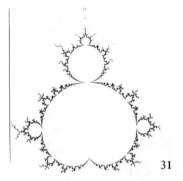

31

This is a fractal snowman who looks like a teenage snowman with major skin and weight issues. If you look carefully, you'll see that what we thought were pimples turn out to be little snowmen that look just like the original, and that each one of the smaller snowmen has smaller snowmen. If you were to magnify any point of the edge of the snowman, you'd see many, many snowmen just like him, no matter how much you magnify the original. In fact, you can magnify off into

122

infinity, and all you will get is endless snowmen, all looking exactly like all the other snowmen.

That's called "self-similarity," and it's one of the things that fractals always do. You can't predict what your fractal is going to look like when you choose the initial values of Z and C above, but you will always get infinite self-similarity. In fact, if you were to try to touch a point on the edge of any fractal, you'd find that you can't actually do it. If you zoom in on any point that you think you have touched, you find empty space surrounded by the same sorts of fractal you have going in all other directions.

Between Dimensions

Curiously, fractals are described as having "fractional dimension." That is, we are used to the first dimension (straight lines), the second dimension (flat planes), the third dimension (space), and the fourth dimension (time). But fractals lie between two and three dimensions, sort of like 2.6 dimensions. Whatever the heck that means.

Beauty

The beauty in fractals is easy to demonstrate. You can see our little snowman in this one, but now he (or she—it's a little hard to tell) seems to be in an actual snowstorm. It's all a fractal, from our simple equation.

And speaking of things that appear in nature, as we continue to look at fractals, we begin to see some similarities between what our little equation produces, and what nature produces. Nautilus shells. Snowflakes. Ferns, like the one below:

32

This fern was produced by using our same simple equation and something called an affine transformation. And we can produce trees. Bushes. Shrubs. Flowers. The stuff of life.

What we clearly see is beauty, complex order, and unpredictability. They are, however, just pictures, not real trees, bushes, shrubs, or flowers. The idea that a simple equation can produce such amazing beauty and complexity is intriguing, though.

In Review

Einstein got rid of the Infinite Universe. QM did serious damage to pure Determinism and pure Reductionism. Now Chaos has let us know that the universe does not run like a big machine, which is to say that nature is inherently unpredictable. Tiny little variations in the starting point ("initial conditions") can cause huge, unpredictable and catastrophic outcomes that can change the direction of world history.

But Chaos also can produce, out of that unpredictability, beauty, order, and possibly something that begins to look a little like life, though we haven't quite figured it out yet. The universe does not run like a big machine, it's not predictable, in fact it's radically unpredictable, so pure Mechanism is dead dead dead.

So off we go into Complexity, where we just might find life life life.

Questions:

1. Are you convinced that there is a natural presence of order in the universe? Why or why not?

2. Is order discovered or imposed by humans on the universe? Defend your answer.

3. Does beauty exist independent of humans, or do humans impose beauty on the universe? Again, why do you think so?

9

Introducing Complexity: Rethinking Change

I need you to stop reading this book and go watch another video. It's not *Jurassic Park*, it's *Jurassic Park: The Lost World*. No, forget it. What you really need to do is to read the book. *The Lost World* is a brilliant Complexity Theory *book*, but the movie version of *The Lost World* was just about a dinosaur in San Diego—that free-range dinosaur we were talking about during a power shortage where people in cars on their way to *eat* dinner got to *be* dinner.

By way of introducing Complexity, here's a clip from *The Lost World* and our old friend, Chaotic Ian:

(Kelly): "One thing I don't understand."

"Only one?" Malcolm said.

"All this business about evolution," she said. "Darwin wrote his book a long time ago, right?"

"Darwin published The Origin of Species in 1859," Malcolm said.

"And by now, everybody believes it, isn't that right?"

"I think it's fair to say that every scientist in the world agrees that evolution is a feature of life on Earth," Malcolm said. "And that we are descended from animal ancestors. Yes."

"Okay," Kelly said. "So, what's the big deal now?"

Malcolm smiled. "The big deal," he said, "is that everybody agrees evolution occurs, but nobody understands how it works. There are big problems with the theory. And more and more scientists are admitting it."[33]

We probably should give a little definition of what Complexity Theory is all about so that you start this chapter confused. Wouldn't want to spoil our record:

A new scientific discipline, called complexity theory, looks at complex systems and their environments in much the same way as chaos theory. George Cowan founded the Santa Fe Institute, in New Mexico, in May, 1984. Stephen Wolfram began the Center for Complex Systems at the University of Illinois, in 1986. Both organizations were founded to investigate complexity. They have defined complexity as "a chaos of behaviors in which the components of the system never quite lock into place, yet never quite dissolve into turbulence either" (Waldrop, 1992, p. 293). [OK, now that's as clear as mud--af.]

The Santa Fe institute is interdisciplinary, making use of economists, physicists, administrators, biologists, and mathematicians. All are working closely together, trying to find order in complex systems.

Complexity lies at the edge of chaos within the fine line that lies between order and chaos. Although this region is thin, it is vast, like the surface of the ocean. The edge of chaos is a transition phase, where life itself is thought to be created and sustained.

Nicolis & Prigogine (1989) . . . define complexity as the ability of a system "to switch between different modes of behavior as the environmental conditions are varied" (p. 218). In other words, complex systems are able to adapt to their environments.[34]

And from another website:

Complexity Theory and Chaos Theory, which is a part of Complexity Theory, have really only taken off with the advent of the computer, which could undertake the massive numbers of calculations required to investigate complex phenomena. It is a new area of science which some people have said could change our lives as much as Michael Faraday's discovery of electricity and its properties. Prof Stephen Hawking has said, "Complexity will be the science of the 21st Century."[35]

Now you know that if Stephen Hawking is talking about it, it's going to be way too hard to understand for, like, humans. It, like, isn't.

We'll get back to Chaotic Ian in a minute—he'll have some interesting tidbits to offer. Let's form an overview of Complexity from some of the stuff from above.

An Overview of Complexity

First, Complexity and Chaos are closely related—as it is above, Complexity is often called the "edge of chaos." That's why we've moved directly into Complexity after the Chaos chapters; they laid the basis for us to continue looking at some great examples.

The definition of Complexity up there is a little thick, so let's unpack it. First, "a chaos of behaviors in which the components of the system never quite lock into place, yet never quite dissolve into turbulence either"—that's the edge of chaos, the place where we haven't quite collapsed into a heap of rubble, but things are moving around quite a bit.

If you want to try to imagine it (and you know something about surfing), it's like the wave after it curls over but before it smashes you into bleeding bits of flesh and fiberglass. There's a lot of controlled chaos in a wave, but clearly there's a lot of order, too, order that allows you to surf along the face, maintaining a delicate balance between art and drowning. When the wave crashes, and if you didn't quite manage to get off the wave before it got onto you, then you experience real chaos: eating sand, fighting the white water to the surface, dribbling salt water out of your nose at embarrassing moments for the next month, wondering how you're going to get back to your car when your swimsuit has left for parts unknown, that kind of thing.

But when you get onto the wave just right, and it's just the right kind of wave, then you experience the creation of something raw and beautiful, art and sport wound up into an adrenaline rush, something gorgeous and transient like a rainbow or a sunset. That's where "the edge of chaos is a transition phase, where life itself is thought to be created and sustained." That's the excitement found in Complexity: we may be at that place where we are discovering where life itself may be created and sustained.

Complexity-Speak

Let's introduce some of the language that is used by Complexity people, as opposed to complex people, who are fun to date, but not to marry:

- *Emergence*
- *Feedback*
- *Self organization*
- *Collective intelligence*
- *Complex Interdependence*
- *Complicated behavior from simple rules/organisms*

Let's let others define them for us, and then we'll define the definitions so that we actually might maybe understand them a little possibly. Yeah, right. These are from Wikipedia.

Emergence *is the process of complex pattern formation from simpler rules. This can be a dynamic process (occurring over time), such as the evolution of the human body over thousands of successive generations; or emergence can happen over disparate size scales, such as the interactions between a great number of neurons producing a human brain capable of thought (even though the constituent neurons are not individually capable of thought). For a phenomenon to be termed emergent it should generally be unpredictable from a lower level description. At the very lowest level, the phenomenon usually does not exist at all or exists only in trace amounts: it is irreducible.*

OK, got all that? Emergence will actually be best understood by looking at some examples, which we'll do in a bit. What it really is is when simple things (like bacteria) somehow work together to solve some problem (Bactine). What emerges is new behavior and new patterns out of the complex relationship that simple things have formed together. When we get to bacteria and slime mold and potted plants later, it'll make more sense.

130

Self-organization—*The most robust and unambiguous examples of self-organizing systems are from physics, where the concept was first noted. Self-organization is also relevant in chemistry, where it has often been taken as being synonymous with self-assembly. The concept of self-organization is central to the description of biological systems, from the subcellular to the ecosystem level. There are also cited examples of "self-organizing" behaviour found in the literature of many other disciplines, both in the natural sciences and the social sciences such as economics or anthropology. Self-organization has also been observed in mathematical systems.*

You may have figured out that there is no definition of self-organization in that definition of self-organization. That should give you a clue that we might not be totally sure what it is, exactly. Good call.

It seems pretty obvious—things get organized themselves without outside help. Duh. That's not the hard part. The hard part is that most of these things shouldn't have the ability to organize themselves, a little bit like your room cleaning itself up while you are downstairs watching *The Simpsons* so that you don't get in trouble with your mom. The organizing seems to come spontaneously out of nowhere, and it generally leads to *emergence*, that is, a new form of order. Simple things get organized spontaneously.

Collective intelligence *is where a large number of cooperating entities can cooperate so closely as to become indistinguishable from a single organism with a single focus of attention and threshold of action.*

Uh, huh. Got it. Here, try this. When bacteria are collectively intelligent, they self-organize into something new and better, and that new and better thing *emerges* as a result of their *collective intelligence* and *self-organization*. Why do I get the feeling that that didn't help too much? We're going to see some great examples of things being collectively intelligent and self-organizing and new things emerging in like, down there on the next couple of pages somewhere.

This is from http://aurora.phys.utk.edu/~senesac/Complexity.html: *The best appraisal of complexity theory that I found came from the book* Complexification *by John L. Casti. In his book, Casti describes some characteristics that separate simple systems from complex systems. He then implies that the closer you look at the details of a simple system, the more complex it becomes. The characteristic differences between simple and complex systems as outlined by Casti are:*

Predictability of behavior - *Simple systems show no surprises, small perturbations to system produces predictable behavior. If we drop a ball it falls to the ground every time. Complex systems however are full of surprises, small perturbations produce unexpected and sometimes counterintuitive behavior. Lower taxes and interest rates lead to higher unemployment.*

Interactions and **feedback/feedforward** loops - *Simple systems have small number of subsystems that interact and less feedback between subsystems. This makes the system less flexible and tends to limit the behavior of the system. The complex system has many dynamically linked subsystems with feedback/feedforward loops that make the system more flexible and allow for a much wider range of behavior. Example: An investor decides to buy a certain stock. The act of buying the stock causes the price of the stock to fluctuate. Other investors notice the fluctuation and decide to buy/sell shares of the stock which also causes fluctuations in price, and so on.*

Decomposability - *Since the components of a simple system interact weakly we can disconnect one or more of the components and the system will behave more or less as it did before. A complex system with its high degree of interaction is very sensitive to these changes. Disconnecting any part of the system will produce drastic changes in the systems behavior.*

Now that we are totally baffled, let's dive in. The summing up of Complexity can be done in a few sentences. They're not hard to understand, but they do tend to get folks riled up.

- The universe is designed to organize itself spontaneously.

- That is, *emergent self-organization* is at the root of existence.

132

- Self-organization results in higher forms of order.

- But the process is unpredictable; you don't know what you'll get until you get it.

A lot of folks don't want anything like self-organization to be bandied around as though it is some sort of scientific principle (more on this in a minute), but as it turns out, it may be (according at least to Dr. Hawking above) the defining principle of scientific inquiry in this century. You can also see the connection with Chaos, remembering fractals and unpredictability.

Questions:

1. How is Complexity Theory similar to Chaos Theory?

2. Talk about fractals in the context of Complexity.

3. Since no one seems to be able to come up with a good definition of self-organization, define self-organization.

10

Introducing Complexity Again

Complexity and Evolution Theory

Back to Chaotic Ian in *The Lost World*:

[Ian Malcolm said,] "The question remained: how does evolution happen? For that, Darwin didn't have a good answer."

"Natural selection," Arby said.

"Yes, that was Darwin's explanation. The environment exerts pressure which favors certain animals, and they breed more often in subsequent generations, and that's how evolution occurs. But as many people realized, natural selection isn't really an explanation. It's just definition: if an animal succeeds, it must have been selected for. But what in the animal is favored? And how does natural selection actually operate? Darwin had no idea. And neither did anyone else for another fifty years."

"But it's genes," Kelly said.

"Okay," Malcolm said. "Fine. We come the twentieth century. Mendel's work with plants is rediscovered. Fischer and Wright do population studies. Pretty soon we know genes control heredity—whatever genes are. Remember, through the first half of the century . . . nobody had any idea what a gene was. After Watson and Crick in 1953, we knew that genes were nucleotides arranged in a double helix. Great. And we knew about mutation. So by the late twentieth century, we have a theory of natural selection which says that mutations arise spontaneously in genes, that the environment favors the mutations that are beneficial, and out of this selection process evolution occurs. It's simple and straightforward. God is not at work. No higher organizing principle involved. In the end, evolution is just the result of a bunch of mutations that either survive or die. Right?

"Right," Arby said.

135

"But there are problems with that idea," Malcolm said. "First of all, there's a time problem. A single bacterium—the earliest form of life—has two thousand enzymes. Scientists have estimated how long it would take to randomly assemble those enzymes from a primordial soup. Estimates run from forty billion years to one hundred billion years. But the Earth is only four billion years old. So, chance alone seems too slow. Particularly since we know bacteria actually appeared only four hundred million years after the Earth began. Life appeared very fast—which is why some scientists have decided life on Earth must be of extraterrestrial origin. Although I think that's just evading the issue."

"Okay . . . "

"Second, there's the coordination problem. If you believe the current theory, then all the wonderful complexity of life is nothing but the accumulation of chance events—a bunch of genetic accidents strung together. Yet when we look closely at animals, it appears as if many elements must have evolved simultaneously. Take bats, which have echolocation—they navigate by sound. To do that, many things must evolve. Bats need a specialized apparatus to make sounds, they need specialized ears to hear echoes, they need specialized brains to interpret the sound, and they need specialized bodies to dive and swoop and catch insects. If all these things don't evolve simultaneously, there's no advantage. And to imagine all these things happen purely by chance is like imagining that a tornado can hit a junkyard and assemble the parts into a working 747 airplane. It's very hard to believe."

"Okay," Thorne said. "I agree."

"Next problem. Evolution doesn't always act like a blind force should. Certain environmental niches don't get filled. Certain plants don't get eaten. And certain animals don't evolve much. Sharks haven't changed for a hundred and sixty million years. Opossums haven't changed since dinosaurs became extinct, sixty-five million years ago. The environments for these animals have changed dramatically, but the animals have remained almost the same. Not exactly the same, but almost. In other words, it appears they haven't responded well to their environment."

"Maybe they're still well adapted," Arby said.

"Maybe. Or maybe there's something else going on that we don't understand."

"Like what?"

"Like other rules that influence the outcome?"

Thorne said, "Are you saying evolution is directed?"

"No," Malcolm said. "That's Creationism and it's wrong. Just plain wrong. But I am saying that natural selection acting on genes is probably not the whole story. It's too simple. Other forces are also at work. The hemoglobin molecule is a protein that is folded like a sandwich around a central iron atom that binds oxygen—like a tiny molecular lung. Now, we know the sequence of amino acids that make up hemoglobin. But we don't know how to fold it. Fortunately, we don't need to know that, because if you make the molecule, it folds all by itself. It organizes itself. And it turns out, again and again, that living things do have a self-organizing quality. Proteins fold. Enzymes interact. Cells arrange themselves to form organs and the organs arrange themselves to form a coherent individual. Individuals organize themselves to make a population. And populations organize themselves to make a coherent biosphere. From complexity theory, we're starting to have a sense of how this self-organization may happen, and what it means. And it implies a major change in how we view evolution."[36]

Before anyone starts getting hostile and throwing things, this section is not about evolution, at least not directly or mainly. It's about Complexity Theory, and one of the things that Complexity has a major impact on could very well be the theory of evolution, but it also has as large an impact on every scientific field, and in fact on every field of human inquiry. As you read above, the Santa Fe Institute, which focuses on the study of Complexity, has "economists, physicists, administrators, biologists, and mathematicians, all working closely together, trying to find order in complex systems." That is, complexity has things to say about economics, physics, administration, biology, and mathematics, as well as everything else that goes on around us.

Yet for people who believe in evolution, complexity theory may mean that something critical is under significant challenge. Randomness and natural selection are under assault by complex order, for reasons

137

contained in the line above: "Complex systems are able to adapt to their environments." Evolution tells us that random mutations that turn out accidentally to be useful in terms of surviving some sort of change in the environment and that can be reproduced in the next generations are what allow an organism or species, completely by accident, to survive. But complexity is telling us that organisms *spontaneously organize in response to changes.* The question then becomes, if it's true, where does this compulsion to organize come from?

Thus, fans of evolution may feel pressure to allow God back into the equation, as many scientists are feeling. For people who believe in God, it may mean that they will have to accept some form of evolution as an operative force in nature. We will not have any final answer to offer in this book, just what we hope is a thoughtful presentation that all will have cause to consider.

Enough introduction. Let's get on with it. We need some real examples that will help us get a grip on what Complexity actually does.

Collective Intelligence, Self-Organization, and Emergence

Let's start with Collective Intelligence, coupled with Self-Organization and Emergence. There are numerous easy examples to find: ants, bees, and bacteria; migrating birds and butterflies (those butterflies again!); and fish, pods of whales, schools of fish. With each one of these (with the exception of the vastly more intelligent whales), singles or small groups of individuals don't act in an especially organized way, apart from doing whatever is needed to survive, mostly eating and hiding from bigger eaters (again, except for whales, who only get eaten by much smaller humans).

But if you get enough ants or bees together, suddenly they begin to do things as a group that they don't attempt as individuals or small groups. Ants make anthills, bees make beehives, bacteria do terrifying things that we'll talk about later, and birds and butterflies migrate in groups for thousands of miles to precise locations. Fish school together for protection or swim upstream to their birthplaces to spawn, and whales cross vast expanses of ocean to return to the same places every year to mate. Ants and bees create hierarchies within their groups, with queens, hunters, gatherers, and workers adopting separate roles in a complex, organized way.

In fact, they accomplish what we are calling *"self-organizing"*—without outside organizational help or any apparent individual intelligence, they do intelligent, organized things as a collective community. They do it by means of feedback—behavioral and biochemical feedback—and what emerges is complex order at a high level. That's *emergence*. There is nothing more natural in nature than organized behavior, and yet there is something almost magical about the beauty of complicated organized behavior from the simplest of organisms. Do you hear the Chaos in that sentence—from simple origins, complicated results?

Philosophical Implications

That leads us to the philosophical implications of Complexity Theory:

• Nature exhibits complex interdependence at the macro and micro levels.

• Everything seems to be related to everything else.

• Thus, we are vastly more than just the sum of our parts.

We'll need to demonstrate further how complexity establishes these, but not before noting that if we are indeed more than the sum of our parts, then Reductionism has bitten the dust, and the Fab Four of determinism, mechanism, reductionism, and infinity are *history*. We could stop now.

It has long been the assumption of science that the universe is a machine made of parts, and that it is nothing more than its parts. But when the parts self-organize according to the subtle rules of nature into something vastly more complex than just the parts, then the

139

machine becomes unpredictable, not like a machine at all, and though we can reduce that machine to its parts, they give no hint as to what potential they have together until they get together.

There's an ad I got from a magazine somewhere advertising something—underwear or feminine hygiene products, probably. It read, "I am a Shell," with a picture of (go ahead, you can guess) a shell. It went on to say, "I can fight cancer. I am *mercenaria mercenaria*. I have an extract in my shell that has the power to slow cancers in mice. I have the power to be the next penicillin. I am more than a shell." [37]

It's neither a bad start nor a bad slogan for all of us: I am more than a shell. I am more than just me-parts. I am an unbelievable number of very small particles all lumped together to become something vastly more than just particles. I am more than a shell.

Bees and Beetles

Here's an excerpt from an article from the *Dallas Morning News* that illustrates the issue of being much more than the sum of just parts in clear, slightly dirty terms: *"One-night stands can be bad news not just for humans, but for other species, too. Bees that are too eager to have sex sometimes try to mate with a clump of beetle larvae, scientists have found."*[38]

Here's what happens. In the Mojave Desert, there are bees and there are blister beetles. The bees have nests, and the beetle larvae need to get to the nests to eat the pollen in order to be able to survive. Beetles are stupid, however, and have their larvae way out in the desert where there are no nests, so the larvae have to figure out a way to get to the nests.

OK, now here's the first problem with this story: If you think beetles are stupid, just imagine how stupid beetle *larvae* are. First, they don't know that nests exist, and second, even if they did, they don't know that the nests have food. These are *larvae*, for crying out loud. They don't know anything at all.

The second problem is that these larvae who shouldn't know anything but seem to, also shouldn't know how to get to the nest, but they do. And how they do it is just baffling, and a little warped.

What they do is hitch a ride on a bee. No big deal, you say. Read more of the article:

To hitch a ride, scientists have found, the larvae arrange themselves into clusters that resemble female bees. When an oversexed male [now that's a redundant phrase -- af] bee lands on them, the larvae crawl onto his stomach and hang on. Then, when the bee gets truly lucky and finds a real female to mate with, the larvae grab onto the female and are carried back to the nest.

Now just picture this: Really stupid beetle larvae, who shouldn't know where they are going or how to get there, somehow figure out *how to make a shape that looks like a female bee!* They group together, they self-organize into precisely the exact shape they need to fool a really stupid, horny, adolescent male bee into thinking he's going to have a good time. Can I just say that *really stupid, horny, adolescent male* is completely redundant, too? I mean, clearly they will mate with anything, unless these larvae are awfully good at pretending to be a female bee.

OK, now really. Beetle larvae, a big group of stupid which-is-to-say-they-literally-have-no-intelligence-whatsoever beetle larvae get together to look like a female bee. And then when the male comes along and has what we could only describe as a singularly unsatisfying sexual experience, they all grab onto the male (with the bee-grabbers that each one of them has thoughtfully been provided by nature).

But they're not done yet! The male bee doesn't go back to the nest. He goes off to find (you can guess) a real female bee, since the first one didn't work out all that well for him, and when he mates with her, she has what we could only describe as a singularly unsatisfying sexual experience, because all the larvae swarm all over her.

So she goes back to the nest to get cleaned up, the larvae get off, and they have lunch.

But they surely don't tell the next group of beetle larvae (whom they crassly give birth to out in the desert, far away from the nest, as though they are trying to keep all the good stuff for themselves) about looking like a female bee and all of that. The next group of larvae has to figure it out for themselves.

As though larvae can figure anything out at all.

It's remarkable and amazing. It's complexity. It's collective intelligence produced somehow via biochemical and perhaps behavioral feedback, though at last reports, we don't yet know how it all works. To say that nature selects for larvae who can figure out how as a group to form something that looks like a female bee, grab onto the male bee for a ride to a real female bee, and ride that female back to her nest for lunch doesn't answer the question of how it works, especially since classic and neo-Darwinian evolution are completely focused on random mutation at the individual level. It just puts essentially meaningless language on the process without unpacking it in any sense. That is, only one beetle larva will have evolved to know how to pretend to be a female bee. That would be useless and therefore nature could not select for the survival of that one beetle larva. It takes a group of larva to form the necessary shape. We need to know what the mechanism is, and the mechanism may be complexity.

Spiders and Wasps

Here's another example, this one of complex interdependence. In the Amazonian rainforest, there is a spider and there is a wasp. The spider spins a web—a nice, normal-looking web that's great for catching bugs.

142

The wasp spots the spider, swoops down, stings the spider, and paralyzes it. The spider's not dead, just asleep. The wasp lays an egg on the spider's abdomen and leaves, never to return.

The spider wakes up and acts like it normally acts: spinning webs—nice, normal-looking webs—catching bugs, and eating them. Then, after seven to ten days, things get different.

Suddenly the spider spins a different kind of web, one with strong cross-beams instead of thin, hard-to-see concentric circles.

The larva hatches, eats the spider, spins a cocoon, and pupates. Nothing that seems all that weird or mysterious, until you stop to think about it for just a second.

Here's what happens. When the wasp stings the spider and injects what might just seem like a paralyzing fluid, it apparently injects within that fluid something . . . else . . . that has the most remarkable impact on the spider. It causes the spider to do several amazing things: First, it waits a week, living life as a normal spider would, spinning normal webs. Second, after waiting a week, suddenly the spider spins a different kind of web, and then gets to be larva food.

The wasp venom *reprograms* the spider's brain to spin a different kind of web, one that is *perfectly designed to support the weight of the cocoon* so that the larva can pupate and survive! It *also* reprograms the spider's brain to wait a week, so that the spider will be well fed, plump and juicy for the larva. The spider needs a normal web for that week to ten days; otherwise, it would starve, since the new web is only good for supporting cocoons, not for catching bugs! In the act of stinging the spider and laying an egg, the wasp manipulates the spider in incredibly specific ways, not just to provide food for the larva, but to provide for a new and specially designed web to support the cocoon.

That's complex. That's sophisticated. These are simple creatures: wasps, spiders, and larva. And yet the relationship of survival (for the wasp) and death (for the spider) is extraordinarily complicated and subtle.

To sum up: somehow the wasp has self-organized in response to the need for the spider to do very specific things. From simple origins, extremely complex and devious results. One could just say that the wasp has evolved to the point to where its venom paralyzes but does not kill the spider and reprograms the spider's brain to wait a week before spinning exactly the right kind of web to host the larvae for dinner. But that doesn't begin to answer the question of what ordering mechanism in nature motivated those changes.

Questions:

1.Bees, ants, fish, butterflies – give some more examples of your own of collective intelligence as you might find it in nature.

2. What are your thoughts about self-organization and emergence?

3. What about the whole being more than the sum of its parts?

11

Complexity, the Brain, and Life Itself

The following excerpt illustrates what we are talking about when we say "complexity." It's from an article called "Bugs in the Brain" in the March 2003 issue of *Scientific American:*

My demoralized insight stemmed from a recent extraordinary paper about how certain parasites control the brain of their host. Most of us know that bacteria, protozoa and viruses have astonishingly sophisticated ways of using animal bodies for their own purposes. They hijack our cells, our energy and our lifestyles so they can thrive. But in many ways, the most dazzling and fiendish thing that such parasites have evolved—and the subject that occupied my musings that day—is their ability to change a host's behavior for their own ends. For instance, certain mites . . . ride on the backs of ants and, by stroking an ant's mouthparts, can trigger a reflex that culminates in the ant's disgorging food for the mite to feed on. A species of pinworm . . . lays eggs on a rodent's skin, the eggs secrete a substance that causes itchiness, the rodent grooms the itchy spot with its teeth, the eggs get ingested in the process, and once inside the rodent they happily hatch.

These behavioral changes are essentially brought about by annoying a host into acting in a way beneficial to the interlopers. But some parasites actually alter the function of the nervous system itself. Sometimes they achieve this change indirectly, by manipulating hormones that affect the nervous system. There are barnacles found in Australia that attach to male sand crabs and secrete a feminizing hormone that induces maternal behavior. The zombified crabs then migrate out to sea with brooding females and make depressions in the sand ideal for dispersing larvae. The males, naturally, won't be releasing any. But the barnacles will. And if a barnacle infects a female crab, it induces the same behavior—after atrophying the female's ovaries, a practice called parasitic castration.

Bizarre as these cases are, at least the organisms stay outside the brain. Yet a few do manage to get inside. These are microscopic ones, mostly viruses rather than relatively gargantuan creatures like mites, pinworms and barnacles. Once one of these tiny parasites is inside the brain, it remains fairly sheltered from immune attack, and it can go to work diverting neural machinery to its own advantage.

The rabies virus is one such parasite. Although the actions of this virus have been recognized for centuries, no one I know of has framed them in the neurobiological manner I'm about to. There are lots of ways rabies could have evolved to move between hosts. The virus didn't have to go anywhere near the brain. It could have devised a trick similar to the one employed by the agents that cause nose colds— namely, to irritate nasal-passage nerve endings, causing the host to sneeze and spritz viral replicates all over, say, the person sitting in front of him or her at the movies. Or the virus could have induced an insatiable desire to lick someone or some animal, thereby passing on virus shed into the saliva. Instead, as we all know, rabies can cause its host to become aggressive so the virus can jump into another host via saliva that gets into the wounds.

Just think about this. Scads of neurobiologists study the neural basis of aggression: the pathways of the brain that are involved, the relevant neurotransmitters, the interactions between genes and environment, modulation by hormones, and so on. Aggression has spawned conferences, doctoral theses, petty academic squabbles, nasty tenure disputes, the works. Yet all along, the rabies virus has "known" just which neurons to infect to make a victim rabid. And as far as I am aware, no neuroscientist has studied rabies specifically to understand the neurobiology of aggression.

Despite how impressive these viral effects are, there is still room for improvement. That is because of the parasite's non-specificity. If you are a rabid animal, you might bite one of the few creatures that rabies does not replicate well in, such as a rabbit. So although the behavioral effects of infecting the brain are quite dazzling, if the parasite's impact is too broad, it can wind up in a dead-end host.

Which brings us to a beautifully specific case of brain control and the paper I mentioned earlier, by Manuel Berdoy and his colleagues at the University of Oxford. Berdoy and his associates study a parasite called Toxoplasma gondii. In a toxoplasmic utopia, life consists of a two-host sequence involving rodents and cats. The protozoan gets ingested by a rodent, in which it forms cysts throughout the body, particularly in the brain. The rodent gets eaten by a cat, in which the toxoplasma organism reproduces. The cat sheds the parasite in its feces, which, in one of those circles of life, is nibbled by rodents. The whole scenario hinges on specificity: cats are the only species in which toxoplasma can sexually reproduce and be shed. Thus, toxoplasma wouldn't want its carrier rodent to get picked off by a hawk or its cat feces ingested by a dung beetle. Mind you, the parasite can infect all sorts of other species; it simply has to wind up in a cat if it wants to spread to a new host.

This potential to infect other species is the reason all those "what to do during pregnancy" books recommend banning the cat and its litter box from the house and warn pregnant women against gardening if there are cats wandering about. If toxoplasma from cat feces gets into a pregnant woman, it can get into the fetus, potentially causing neurological damage. Well-informed pregnant women get skittish around cats. Toxoplasma-infected rodents, however, have the opposite reaction. The parasite's extraordinary trick has been to make rodents lose their skittishness.

All good rodents avoid cats—a behavior ethologists call a fixed-action pattern, in that the rodent doesn't develop the aversion because of trial and error (since there aren't likely to be many opportunities to learn from one's errors around cats). Instead feline phobia is hardwired. And it is accomplished through olfaction in the form of pheromones, the chemical odorant signals that animals release. Rodents instinctually shy away from the smell of a cat—even rodents that have never seen a cat in their lives, rodents that are the descendants of hundreds of generations of lab animals. Except for those infected with toxoplasma. As Berdoy and his group have shown, those rodents selectively lose their aversion to, and fear of, cat pheromones.

Now, this is not some generic case of a parasite messing with the head of the intermediate host and making it scatter-brained and vulnerable. Everything else seems pretty intact in the rodents. The social status of the animal doesn't change in its dominance hierarchy. It is still interested in mating and thus, de facto, in the pheromones of the opposite sex. The infected rodents can still distinguish other odors. They simply don't recoil from cat pheromones. This is flabbergasting. This is akin to someone getting infected with a brain parasite that has no effect whatsoever on the person's thoughts, emotions, SAT scores or television preferences but, to complete its life cycle, generates an irresistible urge to go to the zoo, scale a fence and try to French-kiss the pissiest-looking polar bear. A parasite-induced fatal attraction . . . [39]

Rat Brains

Here we have parasites getting into the brains of rats, causing them to lose their fear of cats, who easily eat the rats, thus giving the parasites a warm and cozy place to reproduce (we're not going to say anything about a cathouse, as it would be impolitic to do so). But they don't just reproduce, they find their way out of the cat in what most of us would consider an unpleasant mode—via the end opposite that of the mouth—whereupon the feces, and the parasites riding in the feces like some sort of stinky subway car, is eaten by . . . rats.

As the writer says, *this is flabbergasting*. Parasites are even stupider than either blister beetles or blister beetle larvae, and yet they are able to manipulate their hosts in very specific ways. The spider continues to weave its web for seven to ten days, and then it prepares both a dinner table and a bedroom for its killer. The rat continues to do all the things that rats do, except that it seems compelled to go find cats. Zombified crabs (as Dave Barry would say, that's a great name for a rock group) act like females (except for zombified female crabs, who act like zombified males acting like females) and go nest, but the nest turns out to be for, you guessed it, larvae, which have arranged it so that neither the males nor the females actually have any role in giving birth to any sort of crab at all.

I'm beginning to think that maybe the larvae are really running things around here.

This article gives stark and fascinating illustrations of one of the key aspects of complexity theory: *Simple organisms solve complex problems in complicated ways, and complexity theory provides the mechanism for change.* That mechanism is the compulsion to order that the universe seems to operate under, and it is a rapid compulsion to order – things tend to happen very quickly as well as unpredictably. One is even tempted to use the word, creatively.

Other intriguing examples abound, including this one about potted plants from Science News Online:

148

Complex computations may also be underway in a bit of office equipment: the potted plant that brightens up the windowsill. Plants may perform what scientists call distributed emergent computation. Unlike traditional computation, in which a central processing unit carries out programs, distributed emergent computation lacks a central controller. Instead, large numbers of simple units interact with each other to achieve complex, large-scale computations. The plants don't add, subtract, multiply, or divide, (but) they do seem to compute solutions to problems of how to coordinate the actions of their cells effectively.[40]

Plants solve problems by coordinating the actions of their cells. Now that's starting to sound like Audrey in "Little House of Horrors." You want to be careful when your plants start coordinating the actions of their cells. They know where you live.

And again from Science News Online, writing about slime molds this time: "Slime molds . . . exist for most of their lives as single-celled, amoeba-like creatures. When their food supply runs dry, *they somehow figure out, through local signals between cells, how to swarm together* into a slug-like, multi-cellular organism that produces the spores that give rise to the next generation."[41]

It's "The Blob" in real time.

And this one will really make you squirm. It's an article from *Wired* magazine's April 2002 issue:

For more than a century, bacterial cells were regarded as single-minded opportunists, little more than efficient machines for self-replication. . . . The sole ambition of a bacterium is to produce two bacteria. [Hear the Mechanism language—the assumption by scientists was that bacteria cells were little machines.]

New research suggests, however, that microbial life is much richer: highly social, intricately networked, and teeming with interaction. . . . Researchers have determined that bacteria communicate using molecules . . . . microbes are able to collectively track changes in their environment, conspire with their own species, build mutually beneficial alliances with other types of bacteria, gain advantages over competitors, and communicate with their hosts—the sort of collective strategizing typically ascribed to bees, ants, and people, not to bacteria.

Their discoveries suggest that the ability to create intricate social networks for mutual benefit was not one of the crowning flourishes in the invention of life. . . .

It was the first.

"Cell-to-cell signaling" communication seems to be the rule, rather than the exception, in every domain of life.[42]

It's a whole new world of science—not tiny bacterial machines, but bacteria that communicate, conspire, and build alliances, and it's not weird or new; it's maybe the first thing that cells learned how to do when they came into being, and it's the norm in biological organisms.

Human Brains

Not only do we seem to have bacteria talking among themselves and parasites, and larva manipulating brains, we have an instance of our brains manipulating themselves. This is from the Associated Press on October 7, 2001:

BRISTOL, Pa.—Christina Santhouse entered high school last month.

Not unusual for most 14-year-olds, but a big deal for a girl living without the right side of her brain. It was removed five years ago after Christina developed a rare, progressive disease that causes uncontrollable seizures.

To be sure, the radical procedure left Christina with serious side effects: She has partial paralysis of her left arm and leg and lost peripheral vision in her left eye. And when she jerks her head the wrong way, she can feel fluid sloshing around the space where part of her brain used to be.

In almost every other respect, though, Christina is a typical teen-ager. Her intellect and memory are fine. She hates algebra, loves 'N Sync. She's got a big "Keep Out" sign on her bedroom door.

Christina's disease, Rasmussen's encephalitis, is an autoimmune disorder that typically strikes children under 10 and may be caused by a virus or immunological reaction.

"It gradually eats away at one hemisphere like a Pac Man, leaving a very debilitated, very handicapped and quite retarded individual," said Dr. John Freeman, one of Christina's doctors and a pediatric neurologist from Johns Hopkins Hospital in Baltimore.

In Christina's case, the disease first appeared as a small tremor in her left foot during a family vacation on the New Jersey shore. An emergency room doctor immediately sent Christina to St. Christopher's Hospital for Children in Philadelphia. After three days of tests, a neurologist delivered the devastating diagnosis.

It was August 1995. Christina was 8 years old.

A month later, Christina went to Johns Hopkins.

While the initial diagnosis had been hard enough to take, Freeman suggested a cure that seemed infinitely worse: He wanted to perform a hemispherectomy, removing the diseased right side of the brain. He said it would stop the seizures.

Christina was wheeled into surgery in February 1996. Fourteen hours later, Christina was in the recovery room.

In the following days, she suffered an intense headache, much worse than any migraine. Yet a month and a half later, she was back in school.[43]

Here's the story—Christina, age 8 or so, had to have half of her brain removed. The whole half. She was having something like 100 seizures every day—about one every fifteen minutes. Half of her brain was dying and taking her with it.

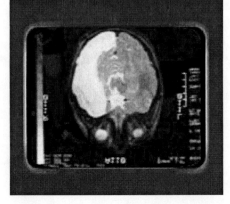

So they took out half of her brain. When she woke up, she was basically normal. She had a little paralysis, a little loss of peripheral vision. But she's a straight-A student, a champion bowler, studying for her black belt in karate. And she has half a brain.

What happened? Her brain (listen for the complexity language here) spontaneously reorganized itself. What emerged was a new and higher form of order, half a brain operating as a full brain. The functions on the right side of the brain, as it gradually died, migrated over to the other side in response to, it might be supposed, biochemical feedback provided by the dying hemisphere. It wasn't a miracle, though it seems miraculous, because it happens all the time in young children. It is more or less normal, miraculous neurological behavior—not even the particles of the brain, but the electrical paths the neurons follow somehow reorganized themselves like a slime mold or a bunch of beetle larvae.

To get a sense of how complicated the brain is, that is, how amazing it is for a brain to reorganize itself spontaneously, it's useful to know how amazing the brain is without being asked to reorganize itself spontaneously. It may either encourage or enrage to be told that you have the most complicated thing in the known universe between your ears.

This is from real scientist Gerald Schroeder's *Hidden Face of God:*

The adult human brain has approximately one hundred billion neurons. So does an infant's at birth. . . . The axon of each neuron connects with as many as a hundred thousand dendrites of other neurons. The branching is stupendous, a million billion connections.

That's 1,000,000,000,000,000 points within our heads.

The brain produces neurons at an astounding rate in the womb. . . . During the nine months of gestation, that averages out to between four and five thousand new nerves each second. These are guided on their journey (to their target organ) by a trillion structural glial cells. Each cell houses . . . nucleus, DNA, mRNA, tRNA, ribosomes, motor proteins, ion channels, and on and on . . . all . . . at the rate of five thousand a second.[44]

The brain is well-accustomed to organizing itself, since the process of producing each brain in every head that ever grew a brain is in itself perhaps the most phenomenally fast and brilliant spontaneous ordering that ever was, all in response to orders delivered by DNA and orchestrated by the delicate protein dance of life.

In fact, that delicate dance is a pretty organized process all of its own, with many elements that classically illustrate complexity—from simple things, complex results, and super-rapid organization at a high level.

Life Itself

Scientists estimate, for example, that as little as 300 genes are needed for life to exist, admitting that there is no real philosophical definition of what "life" or "living" actually are. In the absence of that, they are using the idea of life as being that which can reproduce and respond to its environment.

It doesn't sound like much, but if one goes back to Stanley Miller's primordial soup (the broth he concocted out of the chemicals then thought to have been present on Earth at the time of life's theoretical formation, and the experiment that has long since been discredited), then the mathematics of order gets interesting. Biochemist Fazale Rana writes,

Biophysicist Hubert Yockey has calculated the probability of forming a single gene product as one chance in 1075. Given this probability, Yockey calculated that if the hypothetical primordial soup contained about 1044 amino acids . . . a hundred billion trillion years would yield a 95% chance for random formation of a functional protein only 110 amino acids in length (a single gene product). The universe is about (13.7) billion years old. This means that less than one trillionth of the time has passed that would be needed to make even one of the 250-350 gene products necessary for minimal life, or one of the 1500 gene products necessary for independent life.[45]

153

Biochemist Michael Behe writes of a conversation he and a colleague had one day:

The question was this: "If the first life did arise by random naturalistic processes from a chemical soup, as all textbooks are saying, what exactly are the minimum systems that are required for life?" Together they ticked off a mental list of the minimum requirements: a functioning membrane, a system to build the DNA units, a system to control the copying of DNA, a system for energy processing. Suddenly, they broke off their speculation, looked at each other, and smiled, jointly muttering, "Naaah—too many systems; it couldn't have happened by chance."[46]

An LA Times article on egg fertilization, by David P. Barash, professor of psychology at the University of Washington, demonstrates the power of that observation:

…A particular egg and sperm, each destined to contribute one-half the genome of a future human being, is produced via complex processes of oogenesis and spermatogenesis, respectively. The fated sperm cell migrates through a layer of follicle cells before reaching the egg's "extracellular matrix," known as the zona pellucida. The latter consists of three different glycoproteins, one of which acts as a sperm receptor and binds to its complement on the sperm's head.

This induces a vesicle at the tip of the sperm, the acrosome, to spill its contents of enzymes, which enable the sperm to penetrate the zona and bump up snugly against the egg's plasma membrane. A protein in the sperm's membrane then binds to and fuses with the egg membrane. This in turn triggers depolarization of the latter, which prevents other sperm from entering.

Shortly thereafter, granules in the egg's cortex release enzymes that catalyze additional, long-lasting changes in the zona, achieving a more long-lasting block to other sperm. Pseudopod-like extensions of the egg's interior proceed to transport the sperm into the egg.

If you've been waiting all this time for the genetic fusing of sperm and egg, note that it doesn't happen immediately, at least not in mammals such as ourselves.

Rather, the nuclear envelopes around sperm and egg remain fundamentally distinct through the "fertilized" egg's first mitotic division. Only at this point, with two "daughter" cells already in existence, do the parental chromosomes unite, forming two nuclei. But even at this point, the parental genes remain identifiable and distinct, as either paternally or maternally derived. Paternal and maternal genes thus remain separate for at least 24 hours after sperm successfully breaches those follicle cells, and it takes an additional day or so before their combined influence directs cell function....

LearningKingdom.com says,

Each human cell contains 46 chromosomes, each of which is a DNA molecule. If all the DNA molecules in one cell were unwrapped and stretched out end-to-end, the total length would be about six feet (2 meters). Almost every cell in your body contains six feet of DNA, wrapped up into a very compact space. If all the information in the human genome were printed in small type, it would fill a thousand thick telephone directories.[47]

The *New York Times* tells us, as we read earlier,

The unwound DNA in one person would stretch about 600 million miles—more than six times the distance between the sun and the Earth. . . . DNA may be elegant, but it often has been accorded far greater powers than it possesses. With all the breathless talk of human DNA as a grand epic written in three billion runes, the scientists complain that an essential point is forgotten: DNA, on its own, does nothing. It can't make eyes blue, livers bilious or brains bulging. It holds bare-bones information—suggestions, really—for the construction of the proteins of which all life forms are built, but that's it. DNA can't read those instructions, it can't divide, it can't keep itself clean or sit up properly—proteins that surround it do all those tasks. Stripped of context within the body's cells, DNA is helpless, speechless—DOA. By the same token, cells need their looping lanyards of genes and would grow as dull as hairballs without them.[48]

DNA thus is far more complex, and far less, than we have believed. DNA only works in context, not by itself. It works in a complex system where, once again, the whole is far greater than the sum of the parts, where when you have enough parts, something special and magical happens, perhaps the most special and magical of moments in all existence—life itself.

An Accident or a Design?

A DNA molecule and its home in the cell are enormously interdependent and complex, and yet we have been demanding that this complexity be produced by random accident. In fact, there have only been two choices presented to us—it was random, or it was divine intervention. In the absence of any other mechanism, the fields of battle were thus drawn and the wars raged.

Even and perhaps especially within the confines of evolution itself, the controversy rages. Richard Morris in his superb book entitled *The Evolutionists* details the debate between strict reductionists like Oxford's Richard Dawkins and those with a mind towards finding a complex mechanism to drive evolution, such as the late Stephen Jay Gould of Harvard and Stuart Kauffman of the University of Pennsylvania and the Santa Fe Institute:

Natural selection is more powerful than even advocates of orthodox Darwinism had believed. Scientists have found that, in some species, observable evolutionary change can take place in as little as ten years. Contrary to what had previously been thought, it appears that evolution does not invariably take place through the gradual accumulation of numerous small, favorable mutations. It has been shown that sometimes new species are created when a small number of genetic mutations have large effects. Stephen Jay Gould notes "that Darwin himself had not believed that natural selection was the sole cause of evolutionary change. The 'ultra-Darwinists' or 'Darwinian fundamentalists' shared a conviction that natural selection regulates everything of any importance in evolution"…These evolutionists "push their line with an almost theological fervor."

Paleontologists had adopted the orthodox idea of gradual evolutionary change and had held onto it, even when they discovered fossil evidence to the contrary. They had been trying to interpret fossil evidence in terms of accepted evolutionary ideas. The orthodox Darwinians continued to maintain that…all the observed patterns in evolutionary history were the result of natural selection.

Eldridge and Gould…continued to maintain that the orthodox Darwinists' reductionist methods failed to explain everything. The orthodox view embodied a kind of blindness. The (criticism) that has concerned evolutionary biologists the most has been the charge that (they) often create adaptationist just-so stories that have no basis in fact.

Complex systems have emergent properties. A system of sufficient complexity, whether it is a collection of organisms…or anything else will typically have properties that cannot be explained by breaking the system down into its elements. Complex systems are self-organizing. That is, when a system becomes complex, order will spontaneously appear.

(Stuart) Kauffman speculates that all living systems seek out the edge of chaos. It is this fact that makes life and evolution possible. Initially, Kauffman thought that the genes in a living cell would seek out an edge-of-chaos configuration without the intervention of natural selection. This caused him to wonder if natural selection played any role in evolution at all. Perhaps self-organization could account for everything. Perhaps the appearance of emergent properties in living systems was the most important factor in the evolution of life.[49]

Both sides have the same need: a mechanism of science that produces life in all of its brilliant complexity. If that mechanism allows for the possibility of the existence of an organizing entity, a God, if you like, then the question becomes one of philosophy and personal preference, and no longer science. What we will find, perhaps, is that the evidence for God exists, but not the proof. Once again, for each of us it will come down to a choice made in faith based on the available evidence.

Where twenty-first century science seems to be taking us is into the consideration of complexity theory as that mechanism. Read what the *New York Times* Op/Ed section of March 4, 2003, had to say about it:

Scientists and engineers have learned to harness nonlinear systems, making use of their capacity for self-organization. Lasers, now used everywhere from eye surgery to checkout scanners, rely on trillions of atoms emitting light waves in unison. Super-conductors transmit electrical current without resistance, the byproduct of billions of pairs of electrons marching in lock step. The resulting technology has spawned the world's most sensitive detectors, used by doctors to pinpoint diseased tissues in the brains of epileptics without the need for invasive surgery, and by geologists to locate oil buried deep underground.

But perhaps the most important lesson . . . is how feeble even the best minds are at grasping the dynamics of large, nonlinear systems. Faced with a thicket of interlocking feedback loops, where everything affects everything else, our familiar ways of thinking fall apart. To solve the most important problems of our time, we're going to have to change the way we do science.

For example, cancer will not be cured by biologists working alone. Many cancers, perhaps most of them, involve the derangement of biochemical networks that choreograph the activity of thousands of genes and proteins. As Fermi and his colleagues taught us, a complex system like this can't be understood merely by cataloging its parts and the rules governing their interactions. The nonlinear logic of cancer will be fathomed only through the collaborative efforts of molecular biologists—the heirs to Dr. Watson and Dr. Crick—and mathematicians who specialize in complex systems—the heirs to Fermi, Pasta and Ulam.[50]

We're going to have to change the way we do science. And the ordering power of complexity, the spontaneous reorganizing of organisms, the collective intelligence of unintelligent bits of biology, the cooperative complex interdependence of all manner of organisms, will challenge us to reorganize the way we look at life, the universe, and everything.

Questions:

1. Respond with your thoughts about the statement "Complex systems are able to adapt to their environments."

2. How does Complexity Theory argue against a mechanistic universe? A reductionistic universe?

12

Clumps: A Brief History of Science and God

Ultimately, there is one question that science needs answered, among many others in all directions: Where does the order come from? It is where physics and metaphysics meet, unavoidably and inevitably.

A Primitive God

For centuries, humans answered the unanswered questions of science and nature with the God-of-the-gaps God; anytime we couldn't find an answer to our questions about why nature is the way that it is, why things happen that we can't explain, humans would invoke the gods. In primitive societies, when humans didn't have the answers for the origins of fire, earthquakes, the sun, eclipses, floods, pestilence, plague, famine, anything and everything that lay beyond our ability to reason, the answer was laid at the feet of the gods, and, with the arrival of monotheism, at the feet of God. Where science and human reason failed, God became the answer.

As science progressed over the past few hundred years, the gaps became fewer and fewer. The question of origins is key and pivotal. When Charles Darwin filled that gap, at least with a powerful theory, a fateful assumption was made: *If we filled all the scientific gaps, then there was no longer any role for God to play.* Initially this assumption was made not so much by scientists but by philosophers. (Interestingly, Darwin was a firm believer in God until his young child died, and his anger with God caused him to abandon any form of theistic evolution.)

The Age of Reason

God became intellectually superfluous, the last gasp of the ignorant savage; the weak and pathetic human needs an intellectual crutch to prop up his or her anemic mind. The true intellectual, the truly intelligent human, a man fully capable of achieving the highest levels of reason, no longer has any need of God. Humanity's ability to reason, to observe, to analyze, to break nature into bits and string all the bits together into a cogent and comprehensive whole, that ability has eliminated the need for any gap-God to be invoked. Human's reason itself became all the god that was needed.

The Age of Reason then became the age when humans assumed that their intelligence would solve all of the problems of nature. Not only would they continue to fill in the gaps (and our faith continues to be placed in our ability to solve the unsolved problems), but they would in time eliminate the ills of nature—famine, war, disease, plague, violence, injustice, oppression, racism, bigotry, all of the evils on Earth of human and natural origin would be overcome by our ability to reason and think. Humankind became something like a god, and our hopes in that god were high and utopian.

Rejecting the Existence of God

In light of that, I have to make note of a curious thing. There is a deep, deep flaw in the reasoning above. It looks like this: People of faith made the assumption that when they could not find a natural explanation, only a supernatural answer would fit. God became, at least in large part, that supernatural explanation for what always turned out to be naturally occurring phenomena. People didn't understand the sun, fire, volcanoes, earthquakes, eclipses, or cows, and so they made gods out of these and many things that we now know have a normal, natural, scientific explanation. When people had no scientific explanation for the origin of the universe, then they naturally (or supernaturally) said that God had just waved his magic wand and the universe bounced suddenly into being.

When scientists began to be able to explain, if not everything, at least many of the crucial areas of nature with natural explanations, the supernatural explanations gradually went away. Since God had been defined as He who worked supernaturally, then philosophers rejected *that* God, and in so doing, rejected God *altogether*. We no longer needed supernatural explanations, so we no longer needed God.

They accepted a primitive definition of God as the real definition. They defined God as that thing that was supposed to have made earth, wind and fire, volcanoes, earthquakes, and cows, but since the scientists now knew the scientific causes of all of those things, then, they said, God must not exist. They were as guilty of being primitives as the sun- and fire-worshipers they ridiculed.

For it makes no sense intellectually to say that if we can explain all of nature with good science, then God does not exist. It's a *non sequitur*. If we could explain absolutely all of nature and the universe with a single, simple equation, it doesn't mean that there is or isn't a God. It only means that if God is defined as the thing that does all the things we can't explain, then we have no need of that God, or better, we need to update our definition of God.

It is more and more true all the time: We have no need of a God who fills the gaps in science with supernatural miracles. We are close to filling all of those gaps with good science. But it does not mean that God does not exist, only that we have been operating with the wrong idea about God's role, if he indeed does exist. And since we have found that *supernatural* as a word just means things that happen outside of our understanding of the way nature is, and we have also now found that *natural* things happen outside of our understanding of nature because our understanding of nature didn't include space-time dilation, quantum weirdness, and so on, then the possibility exists for supernatural occurrences that are really natural. *That is, a supernatural*

God has more than enough weird science available to him to do anything naturally, and it'll look supernatural to us casual on-lookers.

Unproven Assumptions

There are still some interesting unproven assumptions nested hidden in the debate, though. First, that we are intellectually capable of filling all the gaps. Quantum Mechanics shows us the limitations of our intelligence, Chaos Theory the limitations of our ability to make predictions, Relativity our limited abilities to experience things that are happening to us. We will never understand the world of the very tiny, and we will never reach far enough back into time to understand all of Big Bang, nor to prove any theory of origins. *Scientific American's* August 1998 issue includes this statement: "The science of the 20th century is showing us, if anything, what is unknowable using the scientific method [and] what is reserved for religious beliefs," argues Mitchell P. Marcus, chairman of computer science at the University of Pennsylvania. "In mathematics and information theory, we can now guarantee that there are truths out there that we cannot find."

Our emotions want to react against those statements, saying, "You don't know that we won't be able to know those things!" and that is both true (I don't) and laden with its own assumption that humans can discover and know everything. That's a powerful and unproven assumption, and modern science seems to suggest strongly that there are physical limitations imposed on us that will continue to prevent access to knowing certain critical things, some of which we'll talk about soon.

Second, even though we may not need supernatural explanations for events that we have discovered have natural origins, it does not mean that God (again, if God exists) could not operate supernaturally if God chose to do so. Indeed, we have already seen that Big Bang was a supernatural event, if not in the way that we traditionally have understood Big Bang, then because nature itself did not exist yet;

162

nature and the laws of physics that drive nature were created at Big Bang. Though whatever caused (if there was a cause) Big Bang may not have been God, it was not a part of nature, since it could not have been a part of something that had not yet been created.

QM also opens the door to the supernatural for us, again, not because we have gaps in nature, but because things seem to be happening outside of our own natural container—perhaps extra-dimensionally (and we operate in a four-dimensional space-time box), perhaps somewhere else, but certainly in a place where we are not allowed to go.

Third, we can be guilty of assuming that humankind's ability to reason represents the highest ability to reason that can exist. Though we might not say it directly, we have acted as though this is the case, certainly in the Age of Reason, if not now, again for reasons that we will discuss in chapter 13. There is a naïve and almost quaint arrogance in that assumption—it is the narcissism of the small child, around whom the universe seems to rotate, the self-centered state of the adolescent, for whom many times nobody else seems to matter. It is almost as though we have decided that although we are not the physical center of the universe, we are the intellectual center.

All we really have to do, of course, is to look back in our human scientific and intellectual history, as we look back on our personal histories, to see how far we have come, in order to imagine how much further we have to go. And clearly, if the universe has been put together by some other entity, there is a level of intelligence so far beyond our own as to be inconceivable, incomprehensible. Perhaps it is our fear that this might be true, that we might be dethroned, that causes us to react so strongly against the idea.

And fourth, that God's existence or non-existence can be established and proven through science, nature, and human reason. Whether we like to admit it or not, there is in nature and in our lives a faith aspect—faith that there are reproducible patterns that help us to

163

establish scientific theory and, subsequently, scientific truth; faith that elevators will not suddenly drop us into the basement; faith that airplanes will take off and land safely; faith that cars will start when the key is turned and will move when put into gear; faith that our choices will work out the way we hope, that our marriages will thrive, our children will love us and do well, our fields of study are right for us and our futures, our jobs will be fulfilling, our lives will be long, healthy and prosperous.

An immediate response to all of that is to say, well, how foolish! There is ample evidence that marriages go bad, that children will disappoint their parents, that jobs and career choices don't work out, that cars refuse to start and that planes crash.

True enough. As Agent Smith asks Neo in their final, climactic fight scene, why, then, do we persist? Because we have evidence that demonstrates a certain (if not high) probability that things will work out well enough, on average. Faith is not blind; it is based on the evidence found in life around us, in history, in experience. And sometimes, we persist even in the face of evidence that logically should cause us to stop. Somehow, we have faith that the best will happen, even if we know that the odds sometimes are against it. Our greatest heroes are those who have persisted in the face of all evidence against it.

That there is faith is at least some evidence that there might in fact be something that asks faith of us.

That there is order all around us, waiting to be discovered, patterns of nature and science that allow us to make predictions into the future and explanations into the past, is at least some evidence that there might be something that has given order to the universe.

And that's what drives the debate in the physics community—there is order, unbelievable, magnificent, subtle, profound, incredibly unlikely order. Where indeed does it come from?

Two Big Bang Surprises

Let's go back to Big Bang for a bit. When the Bang finished its work, here's what we had …

- an empty, expanding, highly ordered universe
- light
- heat
- darkness (after the light faded)
- rules—the laws of physics
  - gravity, electromagnetism, the strong and weak nuclear forces
  - quantum mechanics

Some of these are not surprising. An empty universe clearly without galaxies, solar systems, stars, planets—that would be expected. Yes, the light of the initial super-rapid expansion, along with an immense amount of heat (estimated in the earliest moment of the Bang at 10^{32} degrees), and the darkness and a slow cooling, one that still continues (the heat of the Bang remains with us as the Cosmic Background Radiation at a temperature of 2.7 degrees Kelvin). As unique and amazing as the Bang was, given that it happened, these are to be expected.

But there are two things that surprise: *a highly ordered universe* and *the laws of physics*.

Picture the Bang, if you can. Out of essentially nothing, a super-rapid expansion that gives birth to time, space, and a rapidly expanding universe with nothing in it but what we've mentioned: light and heat. How can it be highly ordered? It seems as though, just like in an explosion, whatever it is that might be there is being scattered willy-nilly in all directions, as disordered as it is possible to be. It's like it was the most violent plane crash in all of history, only ending up with an intact, working airplane instead of a mess, but without the plane. Yes, I know, that makes no sense, but that's what it was.

1. An Ordered Universe

Brian Greene describes it for us in his book *The Fabric of the Cosmos,* a description that centers around the Second Law of Thermodynamics—entropy. The Second Law tells us simply that things fall apart. If you drop an egg on the floor, it falls apart. If you build something out of Legos and come back days, months, years, centuries later, it will have fallen apart. It won't have gotten bigger and more complicated all by itself—left to itself, it will eventually fall apart. *Entropy* is the word physicists use to describe how badly something has fallen apart. Before the plane crash, the plane has "low entropy," that is, it is highly ordered. After the plane crash, it has "high entropy"—it has fallen apart.

Left to itself, the universe will fall apart. Stars will burn out and collapse into black holes or other, less dramatic things. They'll burn their planets up. Galaxies will disappear. Gradually, over time, everything in the universe will move from low entropy (all arranged and ordered and working just fine, thank you) to high entropy (nothing left at all but particles floating in the emptiness of space).

It is, as it happens, a lot more likely for things to be fallen apart than for things to be together. Eggs break when you drop them, but they never put themselves back together. Never. Legos never build themselves into things, but Lego structures, before long, will be just Legos laying around on the floor getting stepped on, swallowed by dogs and babies, vacuumed up, swept into the trash, or thrown back into the (high entropy) box of other Legos.

It is extremely unlikely for things to have low entropy, that is, to be all arranged, ordered, and working very nicely. If you were to travel to Antarctica to go snowmobiling, and out in the snow and ice, miles from anything, you found a big pile of ice shaped exactly like a McDonald's complete with ice burgers, snowmen and women (and snowteens, I guess) standing behind the counter and in front of the snow-fryers, you would say something like, "Oh, my gosh. This is an

166

extremely unlikely low entropy situation here." OK, maybe not, but that's what you would have—something that could not have happened by itself. If some whacko entrepreneur was revealed to have put the McDonald's out there to try to sell burger and fries, then you'd say, "What a lunatic." And if you went back much later, years or decades later, you would expect to find the restaurant in a much higher state of entropy, that is, all fallen apart and collapsed.

Greene looks around at the universe around him, at nature, at humanity, at the incredibly high state of order we find ourselves in, and he is forced to say that this high state of order (or low entropy) is unbelievably unlikely. Even one human brain has a level of order that defies explanation, far less the 60 billion or so human brains that have ever existed on Earth, plus their bodies, their social organization, buildings, roads, philosophies, sciences, technologies, inventions, books, and that's not to mention animals and plants and the ecosystem and the weather and the planets and stars and galaxies and the universe itself.

There are only two possibilities, according to Greene. The universe just popped into existence right at this very second, or it arrived after Big Bang in the highest state of order it would ever be in.

First, the universe as we see it at this moment suddenly has come into existence at this very second, with all of the complexity and order snapping into place in a single instance, our memories of the past created out of thin air, our books of histories coming fully formed into existence with nothing in actual history having really happened.

This case has to be true if the universe started out with high entropy, that is, if the Bang gave us a universe with no order in it, just heat, light, and dust. If the universe started out with very little order, then the Second Law tells us that low order does not produce lots of order, any more than broken eggs jump up off the floor and put themselves back instantly and perfectly into unbroken eggs. In physical terms, entropy always increases, that is, things always get less ordered as time

167

goes on. Things never get more complicated; they only get less complicated, except for women, who constantly get more complicated, which is hard to understand, since all the women I know started pretty complicated in the first place. But in a really good way, I hasten to add.

Thus, the only answer, in this case, is for the universe to have just completely, accidentally and in the face of enormous, unbelievable odds against it just randomly fluctuated into everything we see around us. But in the moment before this one and in the moment after this one, things were a lot less ordered. We are not merely an accidental blip in history, having gotten lucky hundreds of millions of years ago. We got lucky right now, right at this instant.

OK, that's silly, but if the universe started out with low order, there is no way for us to have been produced: higher order never comes from lower order. We could not have come from a less-ordered universe to what we are now. Not possible. Never.

So the only right answer is behind Door Number Two: the universe started out highly, but not quite perfectly, ordered, and has been getting less ordered ever since. We now sit at a lower form of order than yesterday, last week, last month, last year, last century, last millennium, the last million years, or the last billion years.

The universe started out at the lowest state of entropy. It sprang into existence from the Singularity for no physical cause, came into being in a tiny fraction of an instant, expanded to cosmological size in that same instant, and somehow was in the highest state of order that it would ever be in.

OK, I know what you're thinking, 'cause that's what I'm thinking, too: it doesn't look ordered. It looks like a big, empty nothing. Now we need to talk about the Rules.

2. The Laws of Physics

In addition to light, heat, then darkness, and an empty expanding (highly ordered) universe, the Bang also gave us the Rules. The laws of physics.

They came into being along with the Bang. At the Singularity, there no laws of physics, except maybe Quantum Mechanics, but frankly, we'll never know. Here's one of those rules—we'll never be able to know anything further back than what is called Planck Time, which is 10^{-43} second into Big Bang. The laws of physics prevent us from going there. It's a pretty small amount of time, but when you consider that the entire universe came into being in far less than a second, well, 10^{-43} second is still something.

Anyway, the laws of physics came into being with the Bang, and they were pretty basic initially. There was gravity and there was quantum mechanics, along of course with relativity, which is a gravity thing, anyway, and there was electromagnetism and the strong and weak nuclear forces. The potential for all the other rules was there, too, but there wasn't all that much going on for them, so they just hung out for a few billion years until they were needed. We know now that the laws of physics apply everywhere in the universe, not just in our neighborhood, in part because when the laws came into being, all the other neighborhoods in the universe were pretty darn close to ours, since the universe was a lot smaller in those days. Not that they had days, yet. That came later, with the sun, the Earth's rotation, all of that.

Particles materialize. OK, we're getting distracted. There were a lot of things going on in a big hurry in the first few nanoseconds of the Bang, but one of them will concern us at this point. You might remember the article about Big Bang by Gregg Easterbrook in The New Republic (quoted in chapter three). Here's the bit that we want to talk about:

This process unleashed such powerful distortions that, for an instant, the hatchling universe was curved to a surreal degree. Extreme curvature caused normally rare "virtual particles" (quarks--af) to materialize from the quantum netherworld in cornucopian numbers, the stuff of existence being "created virtually out of nothing," as Scientific American once phrased it.

To restate it, the universe was so powerfully curved that Quantum Mechanics kicked in and forced jillions of quarks to spin into existence out of nothingness. This kind of thing happens all the time, and you get a quark and an anti-quark that destroy each other almost as quickly as they come into existence. That's right, you get a little bit of matter and another one of anti-matter, and they wipe each other out. QM tells us that this will happen, and it does. In fact, the particle accelerator in Geneva, Switzerland (called CERN) takes advantage of this all the time. They capture anti-particles along with particles, store them up, and when they get enough of them, they run them around a track that's 26 km long and 100 meters under the ground, accelerate them to nearly the speed of light, and smash them together. I used to live near there and I could hear them screaming. It was ugly.

So we get quarks and anti-quarks all the time, and they nearly always destroy each other. *Nearly* always is the key. QM tells us that there is a one in a billion chance that the quark will escape the anti-quark and float off into space, looking for something to do. I don't know where the anti-quark goes. Anti-Disneyland, I guess.

In the primal moments of the Bang, there were jillions of quarks and anti-quarks being created and destroyed, but every billion-and-first time, the quark would escape and float away.

That's why you and I are here. Without the escaping quarks, no matter, no dirt, no planets, no you, no me, no Walmarts, no nothing.

That's an important rule.

Particles gather into atoms. Here's the next thing: Quarks get lonely. They don't like to hang around with nobody to play with. So quarks are compelled by the laws of physics to combine together to form larger particles, namely protons and neutrons.

170

Protons, neutrons, and electrons really don't like to be alone, either, so they are all compelled by the laws of physical chemistry to combine together to form simple atoms—mostly hydrogen, but also helium, lithium and deuterium (an isotope, a simple variation of hydrogen). If you look back at our table of elements, you'll see that these are the simplest elements (and a very simple isotope) that exist—it's logical that the easiest to form would be those that would form first.

Quantum mechanics produces quarks. Physical chemistry produces simple atoms. What's next?

Atoms clump. Gravity is what's next. The force of Big Bang was enormous, as you might imagine—it takes a lot of oomph to produce a universe out of a tiny speck instantaneously. But the force of the expansion was incredibly carefully balanced. If Big Bang has been too powerful, then all of the particles (the quarks, the subatomic particles and the atoms themselves) would have been blown too far apart to have produced anything.

Anything.

There would have been nothing. Probably not even subatomic particles. Just quarks floating in the vast emptiness of a cooling universe. No dirt. No stars. No planets. No Walmarts.

Nothing.

If the force of Big Bang had been smaller, then the universe would not have lasted long enough to produce anything useful. At one level, there would have been just stars that would not have been able to do what stars eventually did. At another level, the universe would have fallen back in on itself before it produced stars. And there would have been nothing.

The force of the expansion was precise to one part in 10^{60}. That's a one with sixty zeros after it. It looks like this:

1,000,000,000,000,000,000,000,000,000,000,000,000,000,000,000,000,00
0,000,000,000.

If the force of Big Bang had been one part smaller than that, or one part larger, then we get nothing. Not only no life, but no nothing. Either emptiness and dust, or a collapsed universe.

But because the force was just right, and also because quantum physics made the universe slightly uneven, the atoms were close enough together to be brought into clumps by their own electromagnetism and then, as the clumps got larger, by gravity. They were close enough that they rolled down each other's gravitational wells to make clumps.

Some clumps were small and dull, like, I don't know, very short accountants (with apologies both to interesting short people and tall accountants). Others were big enough that the law of gravity caused some other laws to kick in. As gravity drew the atoms together, and if there were enough atoms in one clump, then the pressure of gravity on the particles began to generate heat. Eventually, the clumps began to experience nuclear fusion as the atoms tried harder and harder to push each other out of the way.

The clumps became stars and started to burn. All because of the rules.

Quantum mechanics produce quarks. Physical chemistry produces simple atoms. Electromagnetism and gravity produce clumps, gravity produces heat, and gravity produces stars. The rules gave the universe no choice, as long as things were set up just right.

Conclusion

Now, if you were me or I were you, at this point I would say to me, or you, or whatever, that it seems like the universe is getting more complicated, more ordered, and not less ordered.

So I had to talk to me to help me figure it out. Look at it like this. I used to live in LA and drive long and hideous distances to get, like, anywhere. As I was sitting in traffic, swearing and pounding the seat (which by the way does absolutely no good at making you go faster, just so you know), every now and then I would calm down and start to think.

I thought, you know, if you could get all of the cars to go exactly the same speed and keep exactly the same distance apart, then we could all get to where we want to go without having to sit in traffic until the @$#%&* end of time. Sorry, I'm getting hostile again.

So here's the picture of the perfect freeway: All of the cars are exactly (to an infinite number of decimal points, so we're not saying 100 feet, but 100.000000 . . . to infinity feet) the same distance apart and are traveling at exactly (not just 60 mph, but 60.00000 . . . to infinity mph) the same speed.

That's a highly ordered freeway. That's the highest form of order a freeway can be in. And as soon as it's set up like that, *instantly* it starts to fall apart. Some cars go a little slower, some go a little faster, and in seconds or minutes, the cars are all clumping together and generating heat, because people are getting *hostile* because most of them are stuck behind some *idiot* who's driving too slowly.

That's kinda like what happened to the universe, without the idiots. When it started, it was perfectly ordered, smooth, everything the same in all directions, and as soon as it started, it started to fall apart, and things started clumping together. Quarks clumped into subatomic particles, subatomic particles clumped into atoms, atoms clumped into stars, and eventually, some of the particles clumped together to form you and me and Walmarts.

Though it seems more ordered, and it is in fact more ordered *locally* (that is, where you are), it is less ordered *universally* (that is, everywhere else). It costs the universe energy for you to be a clump, so the energy level goes down in the universe every instant you exist as a clump, and so the universe falls apart a little in order for you to exist.

I'm not completely sure I get it, either, but that's what Brian Greene says, and he's a real scientist, so it must be true.

173

Questions:

1. Is there any limit to the ability of humans to reason, to discover answers to the great questions of life, to discover all there is to know about the universe? Defend your answer.

2. Is faith reasonable? Why or why not?

3. Which is more reasonable to you – a universe that arrived just a moment ago as a result of a random quantum fluctuation, or a universe that arrived 13.7 billion years ago as a result of a random quantum fluctuation? What's the difference?

13

A Pointless Universe: Are We Just a Weed?

So, we still haven't answered the question, Where does the order come from? As Stephen Hawking asks the question, "Why does the universe go to all the bother of existing? What is it that breathes fire into the equations and makes a universe for them to describe?"[51]

The rules kept working. After star birth, which happened very quickly in the history of the universe, within 30 million years of the Bang, the stars were compelled by the nuclear reactions burning hotly within them to cook the heavier elements out of the simple elements that started the reactions. The heavier elements are necessary for life to exist, which clearly could be a problem, since they only existed in the burning stars. They had to be created in the fiery furnace. So how do they get from burning stars to where they need to go?

The stars eventually burn up all their fuel. Wait, you say, if you just get a lump of burned-up star, how do the heavier elements get to the planets to provide the stuff of life? Because when the stars burn up their fuel, the rules compel the catastrophic gravitational collapse that has been teetering on the edge of happening for the life span of the star, the thing that caused the stars to burn in the first place. Gravity takes over when the fuel burns out—the star explodes in a massive nova or supernova that can be as bright as an entire galaxy all by itself. The force of the explosion spreads the heavier elements to the far corners of the universe, including the planets circling other stars. Some folks say that the heavier elements past iron are actually created in the instant of the explosion itself.

So the dust of life was cooked either in the interior or the destruction of stars and spread to Earth (among other places) via huge explosions. Thus, we are each of us a clump of stardust. And it all comes about because the rules forced it to do so. The laws of physics compelled it all to happen. Amazingly, the laws create what seems to be more complex order out of simple starting points (an empty universe, quarks, subatomic particles, atoms, simple elements, and the weakest force in the universe, gravity, plus the other three elementary forces) not only without violating the Second Law, but in fact completely in compliance with the Second Law; the universe is going from higher order to lower order, but in a phenomenal way providing a kitchen that it uses to cook up life.

Figuring the Odds of Order

What are the odds? Is it likely or definite that this empty universe will be ordered? Oxford physicist Roger Penrose, who along with Stephen Hawking made some of the most pivotal discoveries ever made about Singularities and Black Holes, calculated the odds of an ordered universe appearing out of Big Bang. (Note: these are not the odds of our ordered universe, one with galaxies, solar systems, stars, planets, life, people, and the occasional half-caf latte, but a universe with any order at all.)

The odds against an ordered universe are to 10 to the 10 to the 30th power to one, against. We can start to write it down like this: $10^{1,000,000,000,000,000,000,000,000,000,000}$, or 10 to the non-tillionth power. I think.

But if I want to try to write *that* number down, I have a problem. Think of it like this: if I want to write a million down, giving myself a generous 1 second per zero, it takes me six seconds—1,000,000. If I want to count up to a million, one number per second, it would take me eleven days or thereabouts—that's a million seconds. I could write down a number with a million zeros then in about eleven days.

If I want to write down a billion (1,000,000,000 - a thousand million, for our UK friends), it takes me, using the same system, nine seconds. Counting up to a billion, that is, a billion seconds, takes 31.7 years or so. So writing down a number with a billion zeros would take just under 32 years.

If I want to write down a trillion (1,000,000,000,000 - do you Brits call that a billion? no wonder we can't understand each other), a number with 12 zeros, I can do it in 12 seconds. Counting up to a trillion, or writing down a number with a trillion zeros, would take 31,709.8 years, more or less.

So let's just say that I want to write down all of the zeros in 10 to the 10 to the 30th power. I don't want to count up to it; I just want to write the number down.

Six seconds for a million. Nine seconds for a billion. Twelve seconds for a trillion.

For 10 to the 10 to the 30th, I would have to program a computer to write 100 billion zeros a second, and let it run for a million million million times the age of the universe . . . just to write the number down.

That's a very, very big number—the odds against an ordered universe. It is so big that it means that in our present understanding of the order in the universe, it is not possible for this order to have arrived by random chance. If it makes you feel better to say that the odds are very low, then go ahead. But, as physicist (and Big Bang hater) Fred Hoyle said, it is a lot more likely for a tornado to go through a junkyard and assemble a perfect Boeing 747 from the junk on the first try than it is for us to have a universe that is ordered just by chance. Remember the Second Law: Things never get more ordered, just less ordered. The only way that we exist at all is for the universe to have bounced into existence in an extremely, and extremely unlikely, ordered way.

It's a lot more likely for this universe to have life in it than for it to be ordered, and it's really unlikely for life to exist—one chance in 10^{282}, according to one researcher. Now, those odds we can write down— that's about one chance in one million trillion trillion trillion trillion trillion trillion trillion trillion trillion trillion trillion trillion trillion

177

trillion trillion trillion trillion trillion trillion trillion trillion trillion trillion that even one such life-producing and -sustaining planet would occur anywhere in the universe. If you are curious as to how many planets and stars and things there are in the universe, by comparison, it has been calculated that the observable universe has something like 10^{89} elementary particles, many, many, many fewer than 10^{282}; 10^{193}, to be precise. I'd rather not write that down, if you don't mind.

And just to make it worse, Roger Penrose also calculated the odds of life appearing by chance – he figures it at 10 to the 10 to the 123rd. That's a power of four bigger than the odds against an ordered universe. I think I'll pass on trying to write that one down, too.

Scientists Respond

The late Dr. Francis Crick, co-discoverer of DNA, recognized how unlikely it is for life to exist at all on Earth, saying, "an honest man, armed with all the knowledge available to us now, could only state that in some sense, the origin of life appears at the moment to be almost a miracle, so many are the conditions which would have had to have been satisfied to get it going"[52]; to rephrase, life appears to be almost a miracle.

Sir Fred Hoyle, Plumian Professor of Astronomy at Cambridge University, founder of the Institute of Astronomy at Cambridge, and the guy who gave Big Bang its name, said, "The probability of life originating at random is so utterly minuscule as to make the random concept absurd."[53]

As we noted in a previous chapter (it's worth repeating for our purposes here), Dr. Fazale Rana wrote:

Biophysicist Hubert Yockey [which is a name almost as bad as Eugene's and Bertha's back in chapter two] has calculated the probability of forming a single gene product as one chance in 1075. Given this probability, Yockey calculated that if the hypothetical primordial soup contained about 1044 amino acids, a hundred billion

178

trillion years would yield a 95% chance for random formation of a functional protein only 110 amino acids in length (a single gene product). The universe is about (13.7) billion years old. This means that less than one trillionth of the time has passed that would be needed to make even one of the 250-350 gene products necessary for minimal life, or one of the 1500 gene products necessary for independent life.[54]

And from *New Scientist*, September 13, 1997, these words from Marcus Chown:

Why do we live in a universe with three dimensions of space and one of time? According to a Swedish physicist, it is because physics in space-times with more or less than four dimensions would not permit the existence of observers like us. "Such universes would either be too simple, too unstable or too unpredictable," says Max Tegmark of the Institute for Advanced Study in Princeton, New Jersey.

Obviously, life as we know it could not exist in a universe with less than three space dimensions, Tegmark says. What's more, it turns out that the mathematics of any universe with less than three space dimensions forbids the existence of gravity.

There are problems, too, for space-times with more than one time dimension. These could exist while any observer's perception of time could remain one-dimensional. However, this results in a world that is far less predictable than our own, says Tegmark: "The problem is that physics is infinitely sensitive to initial conditions so it's impossible to predict the future." (There's Chaos as applied to physics at the beginning of time.)

Tegmark says that another problem, first pointed out by the Austrian physicist Paul Ehrenfest in 1917, rears its head for space-times with more than three space dimensions: they don't allow anything to exist in a stable orbit. Particles either spiral together or shoot off to infinity. "Not only does this rule out the existence of solar systems but it also rules out the existence of atoms," he says.

So only a universe that leaves three space dimensions and one time dimension unfolded, concludes Tegmark, is likely to provide the richness, predictability and stability to generate interesting structure, including life. "I can't say categorically that other space-times cannot contain observers," he says. "But I'd say the prospects are pretty bleak."[55]

Recognize some of the key elements to existence in the paragraphs above: only a three-dimensional universe can have gravity, and gravity is one of the initial rules that caused everything to happen that led to life's beginning. Also note the Chaos influence—"the problem is that physics is infinitely sensitive to initial conditions so it's impossible to predict the future"—and that clumping doesn't happen in a universe with more than three dimensions.

So as a scientist, here's what you've got: a massively ordered universe that is really unlikely to exist in any ordered state, a universe where time shifts about like a bat after a bug, particles that have definite attitude about being seen, chaotic influences that make it impossible to predict what's going to happen at critical points in history and science, and simple mathematical formulae that seem to aim us in the direction of life itself. Now what do you do with all of that?

The Anthropic Principle

What one highly respected Cambridge physicist named Brandon Carter did was to put it all together in his brain, follow Occam's Razor (that is, that the most obvious and likely answer is probably the real answer), and come up with what he called the Anthropic Principle. It's simple to state and controversial in the extreme: the universe is so incredibly fine-tuned, put together so delicately with so many things in just the right place at the right time in the right way, that the most logical conclusion is that *the universe exists for the purpose of producing humans,* or at least intelligent, thinking beings (that might leave quite a few humans out, but fortunately we only need one or two).

In fact, it's worse than that. Remembering QM and the need for an observer to make an observation in order for reality to exist, physicist John Wheeler in *Cosmic Search Magazine* refines it to become something even more amazing:

We could not even imagine a universe that did not somewhere and for some stretch of time contain observers because the very building materials of the universe are these acts of observer-participancy. You wouldn't have the stuff out of which to build the universe otherwise. This participatory principle takes for its foundation the absolutely central point of the quantum: No elementary phenomenon is a phenomenon until it is an observed (or registered) phenomenon. [That is, *nothing happens without an observation--af*] "What good is a universe without somebody around to look at it?*

The universe only exists because there are observers here to look at it. Without intelligent observers, the universe would be in a quantum state of unreality, not existing in a real sense at all, waiting for an observation to be made. Wheeler goes on to say;

If you want an observer around, you need life, and if you want life, you need heavy elements. To make heavy elements out of hydrogen, you need thermonuclear combustion. To have thermonuclear combustion, you need a time of cooking in a star of several billion years. In order to stretch out several billion years in its time dimension, the universe, according to general relativity, must be several billion years across in its space dimensions.[56]

Why is the universe as old as it is? It needed all that time to produce the things that life is made from, and to produce us from life.

Welcome to one of the most baffling, unanticipated, unexpected, disconcerting conclusions in all of physics, science, metaphysics, philosophy, and theology: According to twentieth-century physics, *the universe had to produce intelligent observers in order to exist*. Which came first, the chicken universe or the egg observers? The universe popped into being out of sheer nothingness, and then hung around in quantum unreality until it could produce something smart to look at it and cause it to exist. Really.

Now, in justifying the Anthropic Principle, we don't really need to talk about the special parameters of physics that make not only the universe but the Earth a very special place indeed; the parameters just make the situation that much more fascinating. Recognize before we

181

get into this that the Principle stands on its own merely on the basis of the need for an observer. The rest, as intriguing as it is, is just icing on the cake.

Icing on the Cake

But what icing it is! As stars burn on the knife's edge of nonexistence, balanced between catastrophic gravitational collapse and nuclear annihilation; so too does the Earth, and in fact humanity, live on the tiniest edge of non-existence. The following few facts from Patrick Glynn will serve as a fine introduction:

• Gravity is roughly 10^{39} times weaker than electromagnetism. If gravity had been 10^{33} times weaker than electromagnetism, "stars would be a billion times less massive and would burn a million times faster."

• The nuclear weak force is 10^{28} times the strength of gravity. Had the weak force been slightly weaker, all the hydrogen in the universe would have been turned to helium (making water impossible, for example).

• A stronger nuclear strong force (by as little as 2 percent) would have prevented the formation of protons—yielding a universe without atoms. Decreasing it by 5 percent would have given us a universe without stars.

• If the difference in mass between a proton and a neutron were not exactly as it is—roughly twice the mass of an electron—then all neutrons would have become protons or vice versa. Say good-bye to chemistry as we know it—and to life.

• The very nature of water—so vital to life—is something of a mystery (a point noticed by one of the forerunners of anthropic reasoning in the nineteenth century, Harvard biologist Lawrence Henderson). Unique amongst the molecules, water is lighter in its solid

182

than liquid form: Ice floats. If it did not, the oceans would freeze from the bottom up and Earth would now be covered with solid ice. This property in turn is traceable to the unique properties of the hydrogen atom. (Those properties are quantum – something called Zero Point Vibrations[1].)

• The synthesis of carbon—the vital core of all organic molecules—on a significant scale involves what scientists view as an astonishing coincidence in the ratio of the strong force to electromagnetism. This ratio makes it possible for carbon-12 to reach an excited state of exactly 7.65 MeV at the temperature typical of the centre of stars, which creates a resonance involving helium-4, beryllium-8, and carbon-12—allowing the necessary binding to take place during a tiny window of opportunity 10^{-17} seconds long.[57] (Atheist and physicist Sir Fred Hoyle used the Anthropic Principle to predict the existence of this "carbon resonance", saying after it was successfully discovered, *"Would you not say to yourself, 'some super-calculating intellect must have designed the properties of the carbon atom, otherwise the chance of my finding such an atom through the blind forces of nature would be utterly miniscule. Of course, you would! A common sense interpretation of the facts suggests that a super-intellect has monkeyed with physics, as well as with chemistry and biology, and that there are no*

---

[1] All the bonds affecting water molecules are ultimately caused by quantum effects, but hydrogen bonds are the result of one of the strangest quantum phenomena: so-called zero-point vibrations. A consequence of Heisenberg's famous uncertainty principle, these constant vibrations are a product of the impossibility of pinning down the total energy of a system with absolute precision at any given moment in time. Even if the universe itself froze over and its temperature plunged to absolute zero, zero-point vibrations would still be going strong, propelled by energy from empty space.

In the case of water, these vibrations stretch the bonds between hydrogen atoms and their host oxygen atoms, enabling them to link up with neighbouring molecules more easily. The result is the highly cohesive liquid that keeps our planet alive. - **Water: The quantum elixir**, 08 April 2006, NewScientist.com news service, Robert Matthews *blind forces worth speaking about in nature. The numbers one calculates from the facts seem to me so overwhelming as to put this conclusion almost beyond question."*)

There are currently thought to be 65 such parameters of existence for life on Earth. At the risk of beating it to death, here are 47 of them:

1. Strong nuclear force constant
2. Weak nuclear force constant
3. Gravitational force constant
4. Electromagnetic force constant
5. Ratio of electromagnetic force constant to gravitational force constant
6. Ratio of proton to electron mass
7. Ratio of number of protons to number of electrons
8. Expansion rate of the universe
9. Mass density of the universe
10. Baryon (proton and neutron) density of the universe
11. Space energy density of the universe
12. Entropy level of the universe
13. Velocity of light
14. Age of the universe
15. Uniformity of radiation
16. Homogeneity of the universe
17. Average distance between galaxies
18. Average distance between stars
19. Average size and distribution of galaxy clusters
20. Fine structure constant
21. Decay rate of protons
22. Ground state energy level for helium-4
23. Carbon-12 to oxygen-16 nuclear energy level ratio
24. Decay rate for beryllium-8
25. Ratio of neutron mass to proton mass
26. Initial excess of nucleons over anti-nucleons
27. Polarity of the water molecule
28. Epoch for hypernova eruptions

29. Number and type of hypernova eruptions
30. Epoch for supernova eruptions
31. Number and types of supernova eruptions
32. Epoch for white dwarf binaries
33. Density of white dwarf binaries
34. Ratio of exotic matter to ordinary matter
35. Number of effective dimensions in the early universe
36. Number of effective dimensions in the present universe
37. Mass of the neutrino
38. Decay rates of exotic mass particles
39. Magnitude of big bang ripples
40. Size of the relativistic dilation factor
41. Magnitude of the Heisenberg uncertainty
42. Quantity of gas deposited into the deep intergalactic medium by the first supernovae
43. Positive nature of cosmic pressures
44. Positive nature of cosmic energy densities
45. Density of quasars
46. Decay rate of cold dark matter particles
47. relative abundances of different exotic mass particles

I know you're dying for me to go through the list one by one in excruciating detail, but forget it. We don't want to lose the point in the midst of all the science, the quantum forest for the quantum trees, as it were. By the way, if a quantum tree falls in a quantum forest, what do you get? Answer's at the end.

Two Sides of the Debate

The point is critical: All scientists, though primarily physicists, look at the information above and much more, and divide themselves neatly into two groups. OK, that's a lie. They all kind of waffle around between the two sides of the debate, some more firmly (but not always) on one side, some on the other, everybody running around furiously trying to figure out how to deal with it. One side says man is an accident and God doesn't exist, and the other side says, well, heck, I don't know, just maybe the Anthropic Principle is (and they just hate to use this word) *true*.

Let's take it one side at a time. One side looks at the Anthropic Principle and the supporting science and decides that, well, let's let them speak for themselves. We'll hear from the other side in the next chapter.

Dr. Steven Weinberg, winner of the Nobel Prize for physics in 1979:

The more the universe seems comprehensible, the more it seems pointless. . . .

At the other end of the spectrum are the opponents of reductionism who are appalled by what they feel to be the bleakness of modern science. To whatever extent they and their world can be reduced to a matter of particles or fields and their interactions, they feel diminished by that knowledge.

I would not try to answer these critics with a pep talk about the beauties of modern science. The reductionist worldview is chilling and impersonal. It has to be accepted as it is, not because we like it, but because that is the way the world works.[58]

Dr. Francis Crick, co-discoverer of DNA:

Your joys and sorrows, your memories and ambitions, your sense of personal identity and free will are in fact no more than the behavior of a vast assembly of nerve cells and their associated molecules.[59]

Can you hear the reductionism in the statements above? Even though QM and complexity have given us a universe that is not *purely* reductionistic, the fact that it is still *relatively* reductionistic still captures some great thinkers like Weinberg and Crick, the latter of whom is clearly no dummy, and I have to tell you, Weinberg is not an idiot either.

Still, their statements are, as Weinberg says, chilling. You are nothing more than particles captured by the vague accidents of physical history. You are nothing more than particles.

Dr. Richard Dawkins, the Oxford zoologist and the pre-eminent voice on the planet for evolutionary theory:

The universe "has precisely the properties we should expect if there is, at bottom, no design, no purpose, no evil and no good, nothing but pointless indifference" . . . human beings are "machines for propagating DNA.[60]

Dawkins reduces humans to hosts for DNA (the title of his pivotal book is *The Selfish Gene,* and the theme is exactly that—DNA uses you to survive) and moves into a completely postmodern direction morally and ethically. Since there is no creator, no god, no being that transcends the universe in any way, then it is only logical—there is no good or evil, no purpose to your existence or mine, no right or wrong, nothing really but, and again it is chilling, pointless indifference. The universe has no concern for the existence of humanity, which is once again just a pointless accident of random physical and biological processes.

Dr. Peter Singer, Princeton professor of bioethics at the University Center for Human Values, takes it to the logical if brutal conclusion in his book *Practical Ethics*:

"Human babies are not born self-aware, or capable of grasping that they exist over time. They are not persons." But animals are self-aware, and therefore, "the life of a newborn is of less value than the life of a pig, a dog, or a chimpanzee." Peter Singer goes on to say that parents should be allowed 28 days after the birth of a severely disabled baby to decide whether to kill it."[61] He writes in a later article, *"In fact I could have left out the word "disabled" altogether. For the reason I have just quickly sketched, I do not think that killing any newborn infant is morally equivalent to killing a rational and self-conscious being."*[62]

It's not fair to react against his thoughts on an emotional level if one accepts the pointless universe stance. Even if you personally are appalled by the apparent inhumanity of the statements above, if the universe is pointless, if there is no good or evil apart from what societies construct around themselves as a way to provide some sort of order to existence, and even if most (though certainly, throughout history, not all) societies would never allow the killing of disabled children, that is just a social convention, an artificial morality that has no absolute transcendent roots or basis. If humans have no value, then baby humans have no value. If the only value that really exists is that of survival of the fittest, then it makes no sense in terms of the survival of the species for us to allow disabled infants to continue to survive. They distract from the survival needs of the rest of us, and if they are allowed to breed and continue the line of disability, then the human race runs the risk of being genetically warped over time.

Remember, survival is all there is in the pointless-universe scenario. Although even survival of the species is not really meaningful in the grand scheme of things, it is at least more comfortable and comforting, and if there is no grand meaning or purpose to existence, then at least we ought to be able to aim for a comfortable life before we are reduced again by death to our particle states. Yes, there might be the excitement of scientific research and discovery, but even that might be considered as or reduced to nothing more than an adrenaline rush for intelligent people in the same way that extreme skiers get a rush from skiing a steep chute, surfers from shooting a major tube, climbers beating a tough pitch, or whatever it is that might give each one of us a bit of energy before we return what's left of our energy to the universe. The most fortunate and gifted among us have something we do that seems to give us purpose, but in a pointless universe, it is a pointless and transient purpose, and quickly we must each leave the stage and return to the bus for that long ride to the next nameless town.

Carl Sagan, in looking at a photo of Earth from a distance of 3.7 billion miles away, had this to say:

Look again at that dot. That's here. That's home. That's us. On it everyone you love, everyone you know, everyone you ever heard of, every human being who ever was, lived out their lives. Our posturings, our imagined self-importance, the delusion that we have some privileged position in the Universe, are challenged by this point of pale light. Our planet is a lonely speck in the great enveloping cosmic dark. In our obscurity, in all this vastness, there is no hint that help will come from elsewhere to save us from ourselves.[63]

Nobel Laureate Jacques Monod said this in 1970, kicking off the God is Dead movement:

The ancient covenant is in pieces. Man knows at last that he is alone in the universe's unfeeling immensity, out of which he emerged only by chance. God has been utterly refuted by science.[64]

The world of science, by and large, welcomed this refutation—it was time to stop giving God credit where none was due and to opening the world of reason to solving all of the problems of the universe that faced us in a logical, systematic way. Thinkers, philosophers, and scientists believed not too long ago that human reason would indeed solve all of the problems of the universe, all the problems of famine, pestilence, plague, war, racism, genocide, lack of potable water, poverty, homelessness and so on. By and large it was thought that God, a.k.a. religion, was more of a problem than a solution, since religious people seemed to be more fond of slaughtering each other than dealing with the survival issues that confronted humanity, and so if reasoning, intelligent people could talk everyone else out of worshipping God and into having a healthy respect for and trust in human reason, then all would be well.

A couple of things happened along the way, though. One of them was World War I, and another was World War II. There was also the Cold War, worldwide famine and starvation, immense global plagues of influenza and other diseases, plus as nations reasoned themselves into industrialization, there was smog, pollution, worker abuse, the threatened destruction of the environment, the hole in the ozone layer, nuclear waste, the Chernobyl and Three Mile Island nuclear power disasters, and a long, long list of things that caused thinkers, philosophers, and scientists to reconsider their hope and belief in humanity and its ability to reason the ecosphere into paradise.

189

It was a subtle shift, not one that got a lot of big press, but it is immensely significant. It is interestingly summarized in a brief scene in *The Matrix*, a movie that takes a strong if not necessarily self-aware stance against it. This quote is from Agent Smith, the computer-generated evil guy-who-looks-like-both-a-bureaucrat-and-everybody's-idea-of-an-FBI-agent, spoken as he is giving the facts of life, the universe, and everything from *The Matrix*'s perspective to Morpheus, a leader of the human resistance movement and a believer in The One:

Every mammal on this planet instinctively develops a natural equilibrium with the surrounding environment, but you humans do not. You move to a location and multiply and multiply, until every natural resource has been consumed. And the only way you can survive is to spread to another area. There's another organism on this planet that follows the same pattern. Do you know what it is? A virus. You human beings are a disease, a cancer of this planet. You are a plague, and we are the cure.[65]

I spoke on Life, the Universe, and Everything in a high school in Kansas once, and just after I finished the five-day series of seminars and was packing myself up in the classroom, one student (who turned out to be one of the very brightest students in the school), looked back over his shoulder and said, as he left, "But humans are just a weed." And he walked out.

We have apparently moved into what one might call "post-humanism." If humanism is a philosophical belief that thoughtful, intelligent, reasoning humans will bring in the promised land of health, wealth, and prosperity to all of humanity, then post-humanism is the belief that humans are destroying the planet and that Earth and all of its creatures would be better off without us.

In truth, if you accept the pointless universe and assumptions about it, this is the logical place to end up. It was clearly never true that human reason was going to solve the world's problems, since it was human reason, by and large, that enabled the world's problems to be so messy. We've done really well at creating, for lack of a better term, weapons of mass destruction in all directions, from nuclear bombs to nuclear waste to greed (which produces sweatshops, economic imperialism, hunger, famine, homelessness, and so on, if not next door, then in the third world) to pollution to . . . well, feel free to make your own list.

In our naïve arrogance, we thought that reason would rule the universe. We've come to find out, in the pointless universe, that reason may do a reasonable job at figuring out how things work, but it does a poor job at controlling nature, especially human nature, and that if the universe is truly pointless, then there's no real point to our continued survival, since we seem to be surviving at the expense of everything else. If survival is in fact the only point in a pointless universe, then humans need to go away to allow the universe to survive better, which it will only do without us, or at least this tiny part of it—earth itself.

In a reductionistic universe, where everything is produced by random chance, where there is no point or purpose to life, no meaning to existence, where survival is the only thing that matters and ultimately even survival has no transcendent meaning, then everything really is just pointless.

But there are some folks who look at the same evidence and see something altogether different.

Questions:

1. Some of the statements and assumptions made about a reductionistic universe, where everything is nothing more than the sum of its parts, use language like "pointless indifference", "no evil and no good", "humans are nothing more than animals", "random chance", "humans are a weed", and so on. What are your reactions and thoughts in this regard?

2. Steven Weinberg makes this statement: *"Though aware that there is nothing in the universe that suggests any purpose for humanity, one way that we can find a purpose is to study the universe by the methods of science, without consoling ourselves with fairy tales about its future, or about our own."* How do you react to that?

3. Is there a useful distinction to be made between religion and the existence of God?

14

The Point: Fire in the Equations

Albert Einstein will start us off with this quote:

The most incomprehensible thing about the universe is that it is so comprehensible.

You might contrast that with the one from Steven Weinberg in the last chapter: *"The more the universe seems comprehensible, the more it seems pointless."*

Weinberg looks at the same universe that Einstein sees, and in fact, he looks at a universe that we understand in many ways only because of Einstein, but he arrives at a far different conclusion. Weinberg sees no point to the universe. Einstein sees that because the universe is comprehensible—because we can understand it; because it has order, rules, the laws of physics; because we can describe it using mathematics; because those laws of mathematics and physics compelled the universe to produce itself out of nothing more than low entropy, gravity, and quantum uncertainty—then there is something deep and mysterious and perhaps wonderful lurking at its core.

The Beautiful Answer

Here's a good story. Einstein took a short amount of time to come up with the Special Theory of Relativity, which is all about Time. I mean, it was like hours, maybe even just on his lunch break, which in Switzerland is two hours, so maybe that was more than enough time to go home, eat, watch a little TV, make out with the wife, discover the Special Theory of Relativity, and be back at work in time to sit on his stool some more.

But as I told you about in chapter three, it took him ten years to polish off the General Theory, which is about gravity, and when he did, he found himself looking for the beautiful answer, or rather, the Beautiful Answer. It was so beautiful that he thought, this can't be right, it's way too cool (I translated that from the Swiss-German myself), so he went in another direction for awhile, finally coming back to the Beautiful Answer, which turned out to be the Right Answer.

This was so impressive that, even now, physicists operate under the principle that the Beautiful Answer is more likely to be the right answer, so if your solution gets too ugly, it's probably wrong. It has become so much a part of physics that some physicists get a little hacked off by it, saying that, hey, maybe the right answer is ugly, OK? I mean, back off, dude!

Where this came into play most impressively in recent times is with the search for the elementary particle. You might recall from your sadly misspent education that one of those old Greek guys figured out that there must be an elementary particle, that is, the building block for everything else that is as small as you can get, and he called it the Atom, after his brother Tom. (Some of that was a lie.)

Not only that, but it was wrong, because the atom is not the smallest particle: atoms are made up of protons, neutrons, and electrons, all of those —ons. So the new theory was that there must be something smaller that makes up all of the —ons. They called it, as we have talked about already, quarks. A quark was supposed to be the smallest possible particle.

So then they found a quark, and everyone was happy for awhile. Then they found another type of quark, but that was OK, too. Then a couple of more quarks. Since they were all quarks (although, really, who can tell? They're very small.), life was cool.

But then they discovered that there might be other elementary particles, and then they starting finding them—neutrinos, gluons, muons, hadrons, leptons, bosons, taus, all kinds of things. A veritable zoo of particles.

194

Here's what they said. This isn't pretty anymore. It can't be right. There must be a more elementary particle than these.

The main reason that they started looking for a different answer was because they no longer had the Beautiful Answer. So it couldn't be the Right Answer.

And so they formulated String Theory, the centerpiece of which is that each particle is formed from a tiny one-dimensional vibrating string, and the different masses (and hence characteristics and properties) of each particle are determined by the rate of vibration of the string. Michio Kaku, one of the founders and Henry Semat Professor in Theoretical Physics at the City University of New York, writes about it like this:

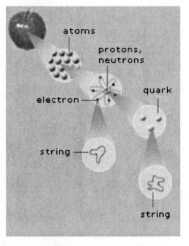

What appeals to me about string theory is that it is gorgeous. The new theory allows the whole of physics to be reduced to the harmonies of these strings. Chemistry emerges as the melodies we play on these strings. The whole universe consists of a symphony of strings, and the "mind of God", about which Einstein wrote so eloquently, is cosmic music resonating through 10 and 11-dimensional hyperspace.

What separates ordinary theoretical physicists from great ones is their ability to discover physical principles and the symmetries lying behind them. Physicists think a theory is beautiful if it can explain the largest amount of physical data with the simplest mathematical structures.

Why does nature use symmetry to express its deepest secrets? No one knows. This is one of the greatest mysteries of all time. I suspect that this is because we are slowly reconstructing the original symmetries that existed at the instant of the big bang, uncovering bits and pieces of new symmetries along the way. If this picture is correct, all the beauty and symmetry we see around us, including sea shells, ice crystals, galaxies, molecules, even sub-atomic particles, are nothing but the pieces of the original symmetry that broke at the instant of the big bang.

Nature, at its most fundamental level, expresses itself in the most beautiful, elegant and symmetrical way.[66]

Why indeed does the universe make sense? Where do the rules come from? As Einstein himself discovered and fostered as perhaps the most powerful method for hunting for solutions in nature, why is it beautiful, and why are the answers beautiful?

To reprise Stephen Hawking's questions: *"Why does the universe go to all the bother of existing? What is it that breathes fire into the equations and makes a universe for them to describe?"* Hawking, like no other person in our times, has a unique ability to phrase the question in precisely the right way. If all that drives the universe is just some mathematical equations, what is it that brings those equations to life? Why is there actually a universe for them to describe? Why does it bother to exist at all?

Why Is There *Anything*?

The fundamental question is, perhaps, this one: Why is there anything at all?

Not why do humans exist? Or, is evolution the answer? Or even, why are the stars the way they are? Why are there galaxies, gravity, gluons, and giraffes?

But, why is there anything at all? Why does our universe even exist in any form? Why is there something rather than nothing?

Even Weinberg is driven from his lonely reductionistic universe to say: *"We may wind up using the Anthropic principle to satisfy our sense of wonder about why things are the way they are."*[67] One hates to say that he seems to be dragged kicking and screaming into the beauty and wonder of nature, but, well, there it is.

And there's another great set of questions to be asked, in no particular order: Why are things beautiful? After all, a sunset is just light, diffracted. Skiing is just friction, delayed and reduced, plus gravity, momentum, fluid dynamics, and occasionally, sudden pain.

Why is a sunset beautiful? Why is skiing fun? Is there an experiential richness to life, the universe, and everything that transcends equations, that manipulates physics into wonder?

Scientists Speak

There are also the Anthropic questions and statements, like this classic one from Freeman Dyson: *"The more I examine the universe and study the details of its architecture, the more evidence I find that the universe in some sense must have known we were coming."*[68]

The universe knew we were coming? What does he mean? He means very much what quantum gravity specialist Paul Davies said:

I belong to a group of scientists who do not subscribe to a conventional religion but nevertheless deny that the universe is a purposeless accident.

Through my scientific work I have come to believe more and more strongly that the physical universe is put together with an ingenuity so astonishing that I cannot accept it merely as a brute fact. There must, it seems to me, be a deeper level of explanation. Whether one wishes to call that deeper level 'God' is a matter of taste and definition.

Furthermore, I have come to the point of view that mind, i.e., conscious awareness of the world, is not a meaningless and incidental quirk of nature, but an absolutely fundamental facet of reality.

That is not to say that we are the purpose for which the universe exists. Far from it. I do, however, believe that we human beings are built into the scheme of things in a very basic way.[69]

Dyson is echoed as well by Stuart Kauffman:

If life in its abundance were bound to arise, not as an incalculably improbable accident, but as an expected fulfillment of the natural order, then we are truly at

197

home in the universe. If I am right, the motto of life is not We the improbable, but We the expected.[70]

Renowned astronomer Alan Sandage joins the chorus nicely:

I find it quite improbable that such order came out of chaos. There has to be some organizing principle. God, to me. . . is the explanation of the miracle of existence, why there is something instead of nothing . . . If God did not exist, science would have to invent Him to explain what it is discovering at its core.[71]

He asks and tries to answer several of our questions—how does order come out of Chaos? Why is there something rather than nothing? What is the organizing principle, that shining vibrancy that arises from chaos into complexity?

Astronomer Alan Dressler phrases it well:

We've abandoned the old belief that humanity is at the physical center of the universe, but must come back to believing we are at the center of meaning.[72]

Humanity at the center of meaning in the universe? The observers it needs to exist? Are humans fundamental to any kind of existence at all? It sounds like science heresy, an outrageous and painful antithesis to post-humanism, pure physics taking us to a place where we don't want to go, not a physical place, but a metaphysical, almost theological place.

Robert Jastrow, founder of NASA's Goddard Space Institute, faces it head-on when he writes:

For the scientist who has lived by his faith in the power of reason, the story ends like a bad dream. He has scaled the mountains of ignorance. He is about to conquer the highest peak. As he pulls himself over the final rock, he is greeted by a band of theologians who have been sitting there for centuries.[73]

And finally, Einstein again, who wrote this in 1950:

Science can only be created by those who are thoroughly imbued with the aspiration toward truth and understanding. The source of feeling, however, springs from the sphere of religion. To this there also belongs the faith in the possibility that the regulations valid for the world of existence are rational, that is, comprehensible to reason. I cannot conceive of a genuine scientist without that profound faith. The situation may be expressed by an image:

Science without religion is lame; religion without science is blind.[74]

I said finally, but I didn't really mean it. Microbiologist and environmentalist René Dubos penned this in 1972:

Progressively, continuously, and almost simultaneously, religious and scientific concepts are ridding themselves of their coarse and local components, reaching higher and higher levels of abstraction and purity. Both the myths of religion and the laws of science, it is now becoming apparent, are not so much descriptions of facts as symbolic expressions of cosmic truths.[75]

I suppose that we need to hear one more time from Hawking, this time finishing his quote:

Although science may solve the problem of how the universe began, it cannot answer the question: Why does the universe go to all the bother of existing? What is it that breathes fire into the equations and makes a universe for them to describe? . . . If we find the answer to that, it would be the ultimate triumph of human reason—for then we would know the mind of God.[76]

What Does It All Mean?

So where does that leave us? We have excellent evidence that these things about the universe are true. The universe is . . .

- finite in space and time (Relativity)
- not purely deterministic (Quantum Mechanics)
- not purely mechanistic (Chaos)
- not purely reductionistic (Quantum Mechanics and Complexity)
- intricately well-ordered and designed to produce order (Complexity)
- here because we observed it? (Anthropic Principle and Quantum Mechanics)

- designed to produce us? (Anthropic Principle)

That allows us to reach some reasonable conclusions:

- There was an "In the beginning."
- We have free will and choice some of the time.
- We will not understand everything.
- We are not machines; there is creative mystery.
- We are vastly more than the sum of our parts.
- The universe is supremely well ordered.
- We seem to matter a great deal to someone.
- Because of this, our lives can have purpose and meaning.

Rather than doing a dance on each one of those bullet points, let's read a brief excerpt from the Jodie Foster movie *Contact*. You might remember that the film was about a scientist named Ellie Arroway who worked for Search for Extraterrestrial Intelligence (SETI, which is a real organization dedicated to establishing contact with alien civilizations). In the course of the film, her team received contact from some sort of intelligent creatures elsewhere in the universe, and those aliens sent the plans for a mysterious machine as a test to see whether or not humans were smart enough to build it.

The machine was built (sparing you the cinematic plot and tension for the moment) though without anyone knowing what it was supposed to do, and Ellie was finally chosen to be the explor-o-naut, the guinea pig who gets to sit in the magic chair to see what happens when the machine is turned on. In the Gemini space program days, she would have been called "spam in a can."

What happens is that her "can" drops a few feet and stops. From the perspective of the on-lookers, nothing of note happened at all.

But Ellie took a relativistic journey via a worm-hole through space-time, an "Einstein-Rosen bridge" that lasted, for her, about eighteen hours, though the Earth-bound watchers only saw a fraction of a second of elapsed time. (Now that you know all about space-time dilation, you'll have no problem with eighteen hours and a fraction of a second being the same.) When she returned, it was to universal (well, not strictly speaking—actually, just everybody on Earth) ridicule and disbelief. She had no evidence (apart from a video recording of eighteen hours of static) of her journey to offer. She was dragged in front of a Senate committee hearing, televised to the entire world, and forced to defend herself. This is her final speech to the committee:

Is it possible that it didn't happen? Yes. As a scientist I must concede that, I must volunteer that . . . I had an experience. I can't prove it. I can't explain it.

But everything that I know as a human being, everything that I am tells me that it was real. I was given something wonderful, something that changed me forever. A vision of the universe that tells us undeniably how tiny and insignificant and how rare and precious we all are. A vision that tells us that we belong to something that is greater than ourselves, that we are not, none of us are alone.

I wish I could share that. I wish that everyone, if even for one moment, could feel that awe and humility and that hope.[77]

It was thoughtful of the screenplay writer to include this little speech for us, because it is brilliantly written. It sums up for us the tension in the debate over origins superbly, giving us clearly the agony and the wonder that scientists feel when looking at twentieth-century science.

Ellie had "a vision of the universe that tells us undeniably how tiny and insignificant" we are. The pointless universe tells us that we are tiny and insignificant, that we have no value or purpose, that there is no meaning to existence. There is no good, no evil, no plan, no creator, nothing but the endless movement of indifferent particles and a blind and instinctive lusting after survival from the particle level to the level of stars, galaxies, and the universe itself. Survival is all that is, and everything is reduced to a mindless machine (even your mind itself) that aims at doing nothing more than reproducing and surviving. Existence, birth, life, death—everything is predetermined and foreordained like balls on a cosmic pool table.

201

"And how rare and precious we all are." But she keeps going, Ellie does. Her journey to the center of the universe has told her that, though we are "undeniably tiny and insignificant," somehow we are rare and precious—"we belong to something that is greater than ourselves." We are more than the sum of our parts, the universe is more than the sum of its parts, we matter because we do not stand alone in a pointless universe, but the universe has a point, because "we are not, none of us are alone."

She looks at this universe with "awe and humility and that hope," awe at the beauty and spectacular wonder at the cosmos. We go back to Paul Davies as he wrote, *"Einstein often spoke of God and expressed a sentiment shared, I believe, by many scientists, including professed atheists. It is a sentiment best described as a reverence for nature and a deep fascination for the natural order of the cosmos."*[78]

And that humility Ellie felt deeply and profoundly, confronted overwhelmingly by the clear knowledge that she, and all of humanity, is tiny and insignificant in the face of the immensity of the universe in size and age and scope.

But finally, that hope that comes from knowing, from being told, from experiencing that we are not merely tiny and insignificant, but rare and precious, and it comes from knowing that we can only be rare and precious if we are rare and precious to something greater than ourselves. We do not give ourselves any transcendent value; that value can only come from outside time and space, outside space-time, outside our amazing little universe.

We have to admit that Ellie is a fictional character, that her journey never happened, that she did not return to tell us that we are rare and precious nor that we belong to something that is greater than ourselves; even if she did, she had virtually no evidence, nothing more than personal experience, and even then, she came back with stories of alien civilizations, not of God.

But there are a couple of things to note. First, SETI is operating under a curious assumption. Its methods are to beam streams of data into space that have recognizable patterns, hoping that intelligent beings will perceive those patterns. Big Bang cosmology has shown us that the laws of physics are the same throughout the universe; thus, intelligent life somewhere else will be operating under those same mathematical laws, and the patterns of mathematics will be able to be seen and recognized. The assumption that SETI is using? That patterns dictate intelligence. If we, or aliens, notice mathematically consistent patterns, then we will assume that intelligent beings are behind them. Randomness means no intelligence. Patterns mean intelligent beings.

It is that same assumption that all scientists who have ever lived have made: there are patterns that we can find in nature that will allow us to understand nature better and to make predictions that will be useful to us. Mathematics is able to be discovered and used only because the patterns of math are consistent and therefore usable. The patterned laws of physics allowed Isaac Newton to predict the orbits of planets around the sun, and Albert Einstein to predict what gravity will do to light.

The patterned laws of chemistry allowed scientists to predict what elements remained to be discovered and how to combine certain chemicals to make others. The patterned laws of biology enabled biologists to understand the biosphere and so much within it. The patterned laws of psychology help therapists understand their patients, and the patterned laws of sociology help researchers to understand societies and cultures. It is only patterns that enable us to do any science at all: science is nothing but the discovering of the patterns that lay waiting to be discovered and used. As Albert said above, all scientists operate under "the faith in the possibility that the regulations valid for the world of existence are rational, that is, comprehensible to reason."

If we human, thinking machines seem to operate generally under the concept that patterns imply intelligence, and if we find order in nature—a highly ordered universe that somehow continues to reorganize itself spontaneously and unpredictably at higher levels of (local) order while still keeping the Second Law intact—it is reasonable, then, to assume that there might be an Orderer someplace. We have vastly more evidence of order in the universe than Ellie Arroway had of aliens at the end of a wormhole. In fact, all of science—ordered, wondrous, magnificent, the crown jewel of human achievement and reason—is evidence of an Orderer outside of our space-time, outside of the Singularity, outside of our own rules of engagement: the rules and laws of physics.

It is not just that there is a suggestion that Big Bang could not have happened without an observation being made, nor that the universe could not exist without observers to Look it into being. It is not just that the parameters of order in the universe are so delicately balanced that the universe seems like an incredibly special place, where any small change in any one of many different numbers of nature would eliminate the order in the universe completely. Nor is it that the chances for life or a life-producing and life-sustaining planet to exist are vanishingly small. A bright scientist might even now be working on discoveries that will show any of these to be false or incomplete. If we try to base our theo-cosmology on one more gap in nature, then when the gap is filled, we are once again faced with the pointless universe.

Ultimately, it is that the universe is highly ordered, that it produces order unpredictably out of chaos, that at its base points it is not predetermined, that gives us our theo-cosmology.

It is not proof. There is no proof. It is, however, reasonable evidence, reasonable enough that the greatest minds alive today struggle with it in the science they pursue day to day. It is once again reasonable and intelligent to believe that *we belong to something that is greater than ourselves.*

Questions:

1. In your opinion, are we "tiny and insignificant," "rare and precious", or somehow both? Justify your thoughts.

2. Albert Einstein said, *"The most incomprehensible thing about the universe is that it is so comprehensible."* Steven Weinberg thinks however that *"The more the universe seems comprehensible, the more it seems pointless."* What do you think?

3. Do humans discover patterns and describe them, or do we impose patterns on a random, patternless universe and use them to describe that universe?

4. Could a rational, intelligent human formulate a personal belief in some sort of God based upon science and nature, or does science mitigate against the existence of God?

P.S. If a quantum tree falls in a quantum forest, it neither does nor does not make a sound unless there is someone there to hear it.

Notes

[1] James Gleick, "Science on the track of God", *The New York Times Magazine*, Jan 4, 1987

[2] http://encyclopedia.thefreedictionary.com/Laplace%27s%20demon

[3] **http://www.churchartworks.com**

[4] www.volftp.mondadori.com

[5] **http://www.churchartworks.com**

[6] Gregg Easterbrook, "Science sees the light", *The New Republic*, Oct. 12, 1998

[7] *USA Today*, February 22, 2000

[8] *Time*, February 24, 2003

[9] www.howstuffworks.com

[10] Neil Gershenfeld and Isaac L. Chuang, www.sciam.com/, "Quantum Computing with Molecules"

[11] *Ibid.*

[12] Mark Buchanon, *Double Jeopardy*, NewScientist, 18 June 2005

[13] Transcribed from *Jurassic Park*, Directed by Steven Spielberg, Written by Michael Crichton

[14] http://www.bbcworld.com/content/clickonline/archive_2001/week45_2001/this week/lorentz.html

[15] http://www.cnn.com/WEATHER/9710/25/plains.snow.pm/index.html

[16] Allen G. Breed, Oct 3, 2002, Associated Press

[17] Marc Spiegler, March 2003, *Wired*, "Avalanche!", emphasis added.

[18] www.nytimes.com/, Aug 25 2003, emphasis added, http://query.nytimes.com/search/query?query=greatest+power+failure&date=past365days&submit.x=8&submit.y=7

[19] AP Sept 29, 2003, emphasis added.

[20] *Rocky Mountain News*, Jan 18, 2003, emphasis added.

[21] Dave Barry, Apr 15 2001, http://www.taxguru.org/invest/davebarry.htm

[22] *New York Times*, www.nytimes.com/indexes/2003/02/25/health/genetics/, emphasis added.

[23] Shannon Jones at http://www.wsws.org/articles/2003/may2003/chal-m06.shtml

[24] *New York Times*, July 22, 2003, Investigators Relive the Shuttle's Demise

[25] www.cs.bsu.edu/homepages/fischer/Journal/01-01/allen.pdf

[26] http://www.mcs.surrey.ac.uk/Personal/R.Knott/Fibonacci/fibnat.html
Dr Ron Knott, Mathematics Department, University of Surrey, UK

[27] www.shodor.org/interactivate/dictionary/f.html

[28] James Gleick, *Chaos: Making a New Science*

[29] *ibid.*
[30] *ibid.*
[31] *ibid.*
[32] *ibid.*
[33] *Jurassic Park: The Lost World*, pp. 205-210
[34] http://www.schuelers.com/ChaosPsyche/part_1_4.htm:
[35] http://complexity.orcon.net.nz/history.html
[36] *The Lost World.* pp. 205-210.
[37] Cisco Systems advertisement, *Wired* magazine, April 2003
[38] This and the following excerpts are from "One-nighter can lead to trouble in the nest," *Dallas Morning News*
[39] Robert Sapolsly, "Bugs in the Brain," *Scientific American,* March 2003, pp.
[40] Science News Online, emphasis added.
[41] Science News Online, emphasis added.
[42] Bonnie Bassler, *Wired* April 2002
[43] Michael Rubinkam, "Girl lives a normal life after losing half of brain: She hates algebra, loves 'N Sync" Associated Press
[44] Gerald Schroeder, *Hidden Face of God*
[45] Fazale Rana. www.reasons.org , and Hubert Yockey, *Information Theory and Molecular Biology* (New York: Cambridge University, 1992), 183, 203-04, 246-57
[46] Michael Behe
[47] LearningKingdom.com
[48] emphasis added.
[49] Richard Morris, *The Evolutionists*, pp 9-10, 111, 127, 135, 184, 227
[50] *New York Times,* March 4, 2003, Op/Ed section, emphasis added.
[51] Stephen Hawking, *A Brief History of Time*
[52] Francis Crick and Leslie Orgel, *Life Itself*
[53] Sir Fred Hoyle, *Evolution From Space*
[54] Fazale Rana
[55] Marcus Chown, *New Scientist,* September 13, 1997
[56] John Wheeler, *Cosmic Search Magazine,* emphasis added.
[57] Patrick Glynn, *God the Evidence*
[58] Stephen Weinberg, *Dreams of a Final Theory*
[59] Francis Crick, *The Astonishing Hypothesis: The Scientific Search for the Soul*
[60] Richard Dawkins, *The Selfish Gene*
[61] http://www.rtlnm.org/newsletters/vivajan00.pdf
[62] Peter Singer, *Severe Impairment and the Beginning of Life,* APA Newsletters, Spring 2000, Volume 99, Number 2
[63] Carl Sagan, *Pale Blue Dot: A Vision of the Human Future in Space*
[64] Jacques Monod, *Chance and Necessity*

[65] Transcribed from *The Matrix*, written and directed by Larry and Andy Wachowski

[66] Michio Kaku, *The Power of Staring*, NewScientist, 16 April 2005

[67] *The New York Times* October 29, 2002, quoted from *A New View of Our Universe: Only One of Many* By Dennis Overbye

[68] Quoted by Paul Davies in *Are We Alone?* among many other places.

[69] Paul Davies *The Mind of God: The Scientific Basis For a Rational World*

[70] Stuart Kauffman, *At Home in the Universe*

[71] Alan Sandage in *The New York Times*

[72] Quoted by Gregg Easterbrook in *Who's Who in the Science and Religion Debate*, http://www.beliefnet.com/story/7/story_784_1.html

[73] Robert Jastrow in *God and the Astronomers*

[74] Gordy Slack, Mother Jones magazine, November/December 1997 Issue, *When Science and Religion Collide or Why Einstein Wasn't an Atheist:* Scientists talk about why they believe in God, emphasis added.

[75] René Dubos, *On Being Human*

[76] Stephen Hawking, *A Brief History of Time*

[77] Transcribed from the movie *Contact*, from the book by Carl Sagan, directed by Robert Zemeckis

[78] Paul Davies from a speech reprinted in *IB World* magazine, December 1997